D1126556

OPTIMALITY IN NONLINEAR PROGRAMMING: A FEASIBLE DIRECTIONS APPROACH

OPTIMALITY IN NONLINEAR PROGRAMMING: A FEASIBLE DIRECTIONS APPROACH

A. BEN-ISRAEL

University of Delaware, Newark, Delaware

A. BEN-TAL

Technion, Haifa, Israel

S. ZLOBEC

McGill University, Montreal, Canada

A WILEY-INTERSCIENCE PUBLICATION

JOHN WILEY & SONS   New York · Chichester · Brisbane · Toronto

Library of Congress Cataloging in Publication Data:

Ben-Israel, Adi.
   Optimality in nonlinear programming.

   (Pure and applied mathematics ISSN 0079-8185)
   "A Wiley-Interscience publication."
   Bibliography: p.
   Includes index.
   1. Nonlinear programming. 2. Mathematical optimization. I. Ben-Tal, A., joint author. II. Zlobec, S., joint
author. III. Title.

T57.8.B46   519.7′6   80-36746
ISBN 0-471-08057-8

Printed in the United States of America

10 9 8 7 6 5 4 3 2 1

To Yoki, Varda, and Inja
our wives

PREFACE

Ignoring the admonishment of an ancient scholar, "Of making books there is no end" (Ecclesiastes XII, 12), we have written yet another book on nonlinear programming, presenting a new unified optimality theory for convex and nonconvex programming with selected applications and computational implementations.

Optimality conditions serve two purposes. The first is to identify optimal solutions in a way that facilitates the study of their relevant properties, such as existence, uniqueness, and sensitivity to data perturbations. The second purpose is to compute optimal solutions, usually by an iterative method employing an optimality condition at each iteration to test the current feasible solution, and to improve it if it is found nonoptimal.

For these purposes it is desirable for optimality conditions to be *constructive* (in the sense that both the objects used and the results obtained are computable) and *widely applicable*. In particular, optimality conditions should be *free of extraneous assumptions* that restrict the applicability and often cloud the issues. They should also aim at providing *complete characterizations* of optimality, that is, they should be both necessary and sufficient.

In terms of these desirable properties our optimality theory is an improvement on the classical optimality theory of Karush-Kuhn-Tucker, which even for convex programming does not give complete characterizations, except under extraneous assumptions (such as constraint qualifications).

Based on constructive notions, such as feasible directions in the convex case and their second-order counterparts in nonconvex programming, our unified approach offers the following features:

1. Complete characterizations of optimality in convex programming (Sections 3 and 4) and a narrowing of the gap between necessary conditions and sufficient conditions in nonconvex programming (Sections 14 and 16).

2. A second-order optimality theory for nonconvex programming in the nondifferentiable case (Section 14) and the differentiable case (Sections

15 and 16). The former results are timely in view of the strong current interest in nondifferentiable optimization. The differentiable results generalize and improve the second-order conditions cited in recent nonlinear programming texts.

3. Feasible direction methods capable of generating directions that are missed by the classical feasible direction methods (such as movements along boundaries) (Chapter 2).

Development of feasible direction methods complemented by a thorough discussion and solution of constrained line search problems (arising in step-size determination). The algorithm given in Section 8 is optimal in a minimax sense, as proved by dynamic programming.

4. Scope of convex programming applications extended to cases where the classical theory may not apply, such as multicriteria problems (Sections 9 and 10) and Chebyshev solutions of nonfeasible programs (Section 11). In particular, Pareto optimal solutions, which are central to economics, are characterized in Section 9.

This book was written with the student and practitioner of nonlinear programming in mind. We tried to avoid unnecessary details, or abstractions that belong in more advanced treatises. All our results are stated for finite-dimensional space; the extension to more abstract optimization problems (e.g., optimal control, semi-infinite programming) is but one of the directions for future research.

A. BEN-ISRAEL
A. BEN-TAL
S. ZLOBEC

October 1980
Newark, Delaware
Haifa, Israel
Montreal, Canada

ACKNOWLEDGMENT

During the development of the results reported here, we benefited greatly from discussions with our colleagues R. Abrams, J. Borwein, L. Kerzner, H. Massam, H. Wolkowicz, and J. Zowe, each of whom has contributed, directly or indirectly, to this book.

We also gratefully acknowledge the help and encouragement of Professors A. Charnes and D. H. Jacobson, as well as support by the University of Delaware, Technion-Israel Institute of Technology, McGill University, National Sciences Foundation, Nasionale Navorsingsinstituut vir Wiskundige Wetenskappe, and Research Council of Canada.

We are thankful to Mrs. P. Andrione for her outstanding typing.

CONTENTS

OPTIMALITY IN NONLINEAR PROGRAMMING: A FEASIBLE DIRECTIONS APPROACH

INTRODUCTION

1 MATHEMATICAL PROGRAMMING: CONVEX AND NONCONVEX

Mathematical programming (hereafter referred to as MP) is the theory, computational methods, and applications of optimization problems stated in terms of

(i) *variables*, x, taken here to be n-dimensional vectors[1] $x = (x_1, x_2, \ldots, x_n)$;

(ii) *the objective function, $f^0 : R^n \to R$,* to be minimized [$f^0(x)$ represents costs or some other undesirable feature]; and

(iii) *constraints*, limiting the variables x to a subset S of R^n, called the *set of feasible solutions* or *feasible set*.

We write the general MP as[2]

$$\min f^0(x)$$

s.t.

$$x \in S.$$

Convex programming (hereafter referred to as CP) is MP where the objective function f^0 is convex, and the feasible set S is a convex subset of R^n, represented here by finitely many convex inequality constraints

$$S \triangleq \{ x \in R^n : f^k(x) \leqslant 0, \quad k \in \mathcal{P} \} \tag{0.1}$$

---

[1] Optimization problems with variables in an infinite-dimensional space, such as problems in the *calculus of variations*, *optimal control*, and *optimization in abstract spaces*, are beyond the scope of this book.

[2] Here "min" stands for "minimize" (or "infimize") and "s.t." means "subject to." Also "MP" is "mathematical programming" or "mathematical program," depending on the context, with a similar ambiguity in the usage of CP, NCP defined below.

1

where $\mathcal{P}$ is a finite index set[3] and $\{f^k : k \in \mathcal{P}\}$ are convex functions: $R^n \to R$.

We recall that a set $C \subset R^n$ is *convex* if $x \in C$, $y \in C$, and $0 < \lambda < 1$ imply $\lambda x + (1 - \lambda)y \in C$. A function $f : C \subset R^n \to R$ is a *convex function* on C if $x \in C$, $y \in C$, and $0 < \lambda < 1$ imply $f(\lambda x + (1 - \lambda)y) \leqslant \lambda f(x) + (1 - \lambda) f(y)$; if the inequality is strict for $x \neq y$, then f is *strictly convex*.

Supported by the powerful structure of convexity (see, for example, ROCKAFELLAR [70a]), CP has a satisfactory and complete theory, and its computational methods are well developed.[4] In comparison, *nonconvex programming* (NCP) is still largely *terra incognita*, devoid of structure and unifying theory.[5]

2   OPTIMALITY AND USABLE DIRECTIONS IN CP

For any feasible solution $x^* \in S$, a *feasible direction for S at x^** is a vector along which a sufficiently small step lies in S (see Definition 1.1). The *set of feasible directions for S at x^** is denoted by $F(S, x^*)$.

For a function $f : R^n \to R$ and a point $x^* \in \text{dom } f$ we denote by $D_f^<(x^*)$ the *set of directions of descent of f at x^**, that is, directions along which the function values initially decrease (Definition 1.2). At any feasible solution x^* a *usable direction* is a feasible direction along which there is an initial decrease in the objective values. The set of *usable directions at x^** is thus $D_{f^0}^<(x^*) \cap F(S, x^*)$.

A fundamental fact of CP is that *a feasible solution x^* is optimal if and only if there are no usable directions at x^**, that is,

$$D_{f^0}^<(x^*) \cap F(S, x^*) = \varnothing, \tag{0.2}$$

see Lemma 3.1. If (0.2) does not hold, then every usable direction is a direction along which the feasible solution x^* can be improved upon, at least initially.

The *feasible direction methods* of CP (ZOUTENDIJK [60] and [76]) are based on this fact and on the possibility of efficiently generating usable directions by solving certain associated LPs (see Section 5).

Moreover, the classical optimality conditions of CP (F. JOHN [48],

---

[3] With an infinite index set $\mathcal{P}$ one gets *semi-infinite CP*, requiring methods more elaborate than those studied here (see, for example, Ex. 1.13).

[4] Particularly for *linear programming* (LP). The efficient algorithms of LP (which are linear algebraic in nature, and make little use of convexity per se) are used in many iterative methods of CP; see, for example, Section 5.

[5] Except where convexity crops up, as in problems featuring mild generalizations of convexity (see, for example, MANGASARIAN [69] and AVRIEL [76]) that are still covered by CP theory, or in algorithms based on "convexification," for instance, augmented Lagrangian methods (HESTENES [75]).

KUHN-TUCKER [51]), as well as many of their extensions and generalizations, are obtained by representing usable directions in terms of equations
and inequalities in the gradients of the functions f^0, $\{f^k : k \in \mathcal{P}\}$.

3 OPTIMALITY, DIRECTIONS OF DESCENT, AND DIRECTIONS OF CONSTANCY IN CP

The usefulness of the optimality condition (0.2) depends on whether it is
possible to conveniently represent the sets $D_{f^0}^{\lessgtr}(x^*)$ and $F(S, x^*)$, or to
check that their intersection is empty.

The set $D_{f^0}^{\lessgtr}(x^*)$ raises no problem, since the set of descent directions of a
convex function f is given in terms of the directional derivatives of f as
follows (see Lemma 1.4):

$$D_f^{<}(x^*) = \{d : f'(x^*, d) < 0\} \tag{0.3}$$

or, if f is differentiable,

$$D_f^{<}(x^*) = \{d : \nabla f(x^*)^T d < 0\}. \tag{0.4}$$

For the set $F(S, x^*)$ we have the representation (see Theorem 1.7)

$$F(S, x^*) = \bigcap_{k \in \mathcal{P}(x^*)} D_{f^k}^{\lessgtr}(x^*) \tag{0.5}$$

where $\mathcal{P}(x^*)$ is the *index set of binding constraints at* x^* [see (1.16)] and
$D_f^{<}(x^*)$ denotes the *set of directions of nonascent of* f *at* x^* (Definition 1.1),
which for a convex function f can be written as [see Lemma 1.3b]

$$D_f^{\lessgtr}(x^*) = D_f^{<}(x^*) \cup D_f^{=}(x^*) \tag{0.6}$$

where $D_f^{=}(x^*)$ denotes the *set of directions of constancy of* f *at* x^*. Combining (0.2), (0.5), and (0.6), a necessary and sufficient condition for a feasible
solution x^* to be optimal is that for every subset Ω of $\mathcal{P}(x^*)$ the following
intersection $D(\Omega, x^*)$ is empty (see Theorem 3.2):

$$D(\Omega, x^*) \underline{\underline{\Delta}} D_{f^0}^{\lessgtr}(x^*) \cap D_{\mathcal{P}(x^*)\backslash\Omega}^{<}(x^*) \cap D_{\Omega}^{=}(x^*) = \varnothing \tag{0.7}$$

where

$$D_{\mathcal{P}(x^*)\backslash\Omega}^{<}(x^*) = \bigcap_{\substack{k \in \mathcal{P}(x^*) \\ k \notin \Omega}} D_{f^k}^{<}(x^*)$$

$$D_{\Omega}^{=}(x^*) = \bigcap_{k \in \Omega} D_{f^k}^{=}(x^*).$$

There are two apparent difficulties with this approach:

(i) *Complexity.* One has to check all subsets Ω of $\mathscr{P}(x^*)$ in (0.7).[6]

(ii) *Constancy directions.* While the set of descent directions of a convex function f has useful representations [see (0.3), (0.4)], the set of constancy directions can be quite troublesome (see, for example, Exs. 1.7 and 1.8)[7].

4 PRIMAL AND DUAL OPTIMALITY CONDITIONS

The optimality conditions (0.2) and (0.7) are examples of *primal conditions*, stated in terms of feasible solutions x^* and various directions at x^*, which are elements of the same space, the *primal space*.

Certain primal optimality conditions [such as (0.2) and, in the differentiable case, (0.7)] contain a statement about convex sets having an empty intersection, a statement that, by use of an appropriate *separation theorem* (such as one of the theorems of the alternative; see, for example, MANGASARIAN [69]), can be restated as a *dual optimality condition*, involving elements of the *dual space*,[8] for example dual sets and multipliers. Examples of dual optimality conditions are the classical conditions of F. John and Kuhn-Tucker.

The separation theorems used in this book are the Dubovitskii-Milyutin Theorem (Theorem 2.3) and its generalization (Theorem 14.1), translating primal optimality conditions of CP and NCP, respectively, into dual conditions.

5 THE CLASSICAL OPTIMALITY CONDITIONS

The two difficulties mentioned in Section 0.3 above can be avoided simultaneously by considering only the empty set $\Omega = \varnothing$ in (0.7). This leads to the following necessary condition for the optimality of x^*:

$$D_{\hat{f}}(x^*) \cap D_{\hat{\mathscr{P}}(x^*)}^{\leq}(x^*) = \varnothing \tag{0.8}$$

whose dual statement (in the differentiable case) is the necessary condition for optimality of F. JOHN [48] (see Table 3.1).

The set of nonascent directions of a differentiable convex function f can

---

[6]This is not necessarily a disadvantage in computational implementation, since some Ω may yield usable directions of steeper descent than the single $\Omega = \varnothing$ considered in classical theory (see Section 6).

[7]Which explains why constancy directions were avoided in the MP literature before they were introduced in BEN-TAL, BEN-ISRAEL, and ZLOBEC [76].

[8]In this book all spaces are finite dimensional, so there is no need to distinguish between a vector space and its dual.

also be approximated by

$$\left\{ d : \nabla f(x^*)^T d \leqslant 0 \right\}.^9 \qquad (0.9)$$

Using (0.2) and (0.5), one then obtains the optimality condition of KUHN-TUCKER [51] (see Table 3.1), which is sufficient, but not necessary, unless certain additional hypotheses (*regularity conditions* or *constraint qualifications*[10]) are satisfied. Owing to its simplicity, analogy with the classical Lagrange condition (for optimization under equality constraints), important computational applications,[11] and economic interpretations,[12] the Kuhn-Tucker optimality condition[13] has become the cornerstone of *nonlinear programming* (NLP).[14]

6   AN ALTERNATIVE APPROACH

In a sufficiently general case (including most CPs of practical interest), the two difficulties mentioned in Section 0.3 above can be overcome as follows:

(i) *Complexity.*   Actually one needs to check only one subset of $\mathcal{P}(x^*)$ in (0.7), namely, the *minimal index set of binding constraints*[15]

$$\mathcal{P}^= \underset{\Delta}{=} \left\{ k \in \mathcal{P} : x \in S \Rightarrow f^k(x) = 0 \right\} \qquad (0.10)$$

   (see Theorem 3.6). The set $\mathcal{P}^=$ still needs to be computed (see Algorithm 3.9 and Exs. 3.6 and 3.7).

(ii) *Constancy directions.* For the important special case of *faithfully convex functions*[16] (ROCKAFELLAR [70b]), the set of constancy directions $D_f^=(x^*)$ is a fixed subspace, independent of the point x^*. An algorithm for computing $D_f^=(x^*)$ is given in WOLKOWICZ [78] (see Exs. 1.11 and 1.12).

---

[9] In general, the set (0.9) properly contains $D_f^\leqslant(x^*)$ (see Lemma 1.5c).

[10] The best known of which is the *Slater condition*, which states that there exists an $\bar{x} \in S$ with $f^k(\bar{x}) < 0$ for all $k \in \mathcal{P}$.

[11] See, for example, ZOUTENDIJK [76, Section 15.7].

[12] See, for example, ROCKAFELLAR [70a, Section 28].

[13] Or the *Karush-Kuhn-Tucker condition*; see KUHN [76] and KARUSH [39].

[14] Which explains the extensive research on constraint qualifications (see, for example, the survey in PETERSON [73]), needed for the Kuhn-Tucker condition.

[15] The set $\mathcal{P}^=$ is empty if and only if Slater's condition holds, a hypothesis under which the Kuhn-Tucker optimality condition is also necessary.

[16] Which includes the analytic convex functions, and therefore most convex functions of practical interest.

7   THE ESSENCE OF THIS BOOK

This book presents a unified approach to both CP and NCP based on the constructive[17] notion of directions,[18] yielding complete characterizations of optimality[19] and efficient feasible direction methods in CP. In the case of NCP, it describes necessary conditions and sufficient conditions, which generalize and extend known results.

8   THE INTENDED READER AND THE NECESSARY
MATHEMATICAL BACKGROUND AND PREPARATION

This book is intended for students and practitioners of MP, be they mathematicians, operations researchers, engineers, or others. It can be used for self-study or as a textbook in a one-semester graduate course.

The reader is assumed to have a knowledge of linear algebra and of calculus of several variables, such as would normally be gained by completing introductory courses in these subjects. The reader should also be at least superficially familiar with the rudimentary properties of convex sets and convex functions in R^n.

9   AN OUTLINE

The book is divided into two parts, (I) *Convex Programming* and (II) *Nonconvex Programming*. The first part (Chapters 1–3) deals chiefly with the ramifications and applications of the fact (0.2) that in CP, optimality is equivalent to the absence of usable directions. The resulting characterizations of optimality have *primal versions* and *dual versions*. The primal versions are stated in terms of elements of the solution space: feasible directions, descent directions, and constancy directions, studied in Section 1. The dual versions use separation theorems (such as the Dubovitskii-Milyutin Theorem 2.3), which characterize empty intersections [such as (0.2) and (0.7)] in terms of elements of the dual space: dual sets and separating hyperplanes, studied in Section 2.

Section 3 develops the main primal and dual characterizations of optimality in CP, in particular characterizations based on the minimal index set of binding constraints $\mathcal{P}^=$ (studied in Lemma 3.5 and Algorithm 3.9). The classical optimality conditions of F. John and Kuhn-Tucker are obtained as special cases (see Table 3.1).

---

[17] Constructive in the sense of computable.

[18] First-order directions for CP (see Section 1) and second-order directions for NCP (see Section 12).

[19] Requiring no constraint qualification.

A parametric approach to the representation of nonascent directions (Lemma 4.1) leads to new, alternative characterizations of optimality studied in Section 4. The parametric characterizations are especially important in feasible direction methods for CP, studied in Chapter 2.

The *method of feasible directions* is an iterative method

$$x_{\text{NEW}} = x_{\text{OLD}} + \alpha^* \bar{d} \qquad (0.11)$$

where $\bar{d}$ is a usable direction at x_{OLD}, and α^*, the *step size*, is determined by the (single variable) constrained optimization problem

$$\min \left\{ f^0(x_{\text{OLD}} + \alpha \bar{d}) : \alpha \geqslant 0, \qquad x_{\text{OLD}} + \alpha \bar{d} \in S \right\}.$$

In Zoutendijk's classical method (ZOUTENDIJK [60], [76]) the usable direction $\bar{d}$ is generated by solving the LP associated with the F. John condition (0.8) [i.e., (0.7) with $\Omega = \emptyset$]. The directions so generated tend to point into the feasible set,[20] resulting in the computational disadvantage discussed in Section 5. By considering other subsets Ω in (0.7), and the parametric optimality conditions of Section 4, one is led to the modified direction generators of Section 6 and to the *parametric feasible direction* (PFD) algorithm of Section 7, which yield usable directions of steeper descent and other advantages over the classical methods.

In Section 8 we present a constrained line search method for determining the optimal step size α^* in (0.11), based on GAL, BACHELIS, and BEN-TAL [78].

Selected applications of CP where the Slater condition cannot hold (and consequently the Kuhn-Tucker condition generally fails) are discussed in Chapter 3. These applications include Pareto optimality in CP (Section 9), lexicographic multicriteria optimization (Section 10), and the Chebyshev solution of CP (Section 11).

Necessary conditions for general NCP are studied in Chapter 4. The required primal elements are *second-order directions of decrease* and *tangent directions*, introduced in Section 12. The appropriate dual elements studied in Section 13 are *support functions* of subsets S of R^n, $\delta^*(\cdot \mid S): R^n \to R$, defined by

$$\delta^*(y \mid S) \underline{\Delta} \sup\{ y^T x : x \in S \}$$

with effective domain $\Delta(S)$. A necessary condition for optimality in NCP is the nonexistence of improving directions, which is interpreted geometrically to mean that certain convex sets have an empty intersection. This consideration calls for a separation theorem, and the one used here is the following generalization of the Dubovitskii-Milyutin theorem.

---

[20] Since they are directions of descent for all the binding constraints.

Theorem 14.1. Let S_0 be a convex set and let $S_1, S_2, \ldots, S_l$ be open convex sets. Then

$$\bigcap_{k=0}^{l} S_k = \varnothing$$

if and only if there exist vectors

$$y^k \in \Delta(S_k), \qquad k = 0, 1, \ldots, l,$$

not all zero, such that

$$\sum_{k=0}^{l} y^k = 0 \tag{0.12}$$

$$\sum_{k=0}^{l} \delta^*(y^k \mid S_k) \leqslant 0. \tag{0.13}$$

In applying Theorem 14.1 to obtain necessary optimality conditions for NCP, (0.12) corresponds to the Euler-Lagrange condition, and the inequality (0.13) results in the Legendre inequality (see Theorem 14.2).

The differentiable case of NCP, studied in Chapter 5, gives rise to simplified necessary conditions (Section 15) and sufficient conditions (Section 16). The *differentiable nonlinear program* (DNLP) chosen to illustrate our results is

(DNLP)

$$\min f^0(x)$$

s.t.

$$f^i(x) \leqslant 0 \qquad i \in I \underline{\Delta} \{1, \ldots, m\}$$

$$h^j(x) = 0 \qquad j \in J \underline{\Delta} \{1, \ldots, p\}$$

where f^0, $\{f^i : i \in I\}$, $\{h^j : j \in J\}$ are twice continuously differentiable functions: $R^n \to R$. The *Lagrangian* function $L : R^n \times R^{m+1} \times R^p \to R$ for (DNLP) is

$$L(x, \lambda, \mu) \underline{\Delta} \sum_{i=0}^{m} \lambda_i f^i(x) + \sum_{j=1}^{p} \mu_j h^j(x), \tag{0.14}$$

and for a feasible solution x^* we denote

$$D(x^*) \underline{\Delta} \left\{ d : \begin{array}{ll} \nabla f^i(x^*)^T d \leqslant 0, & i \in I_0(x^*) \\ \\ \nabla h^j(x^*)^T d = 0, & j \in J \end{array} \right\} \tag{0.15}$$

where

$$I_0(x^*) \underset{=}{\Delta} \{0\} \cup I(x^*).$$

A distinguishing feature of our theory is that the multipliers λ, μ depend on the direction d, as illustrated by the following two sample results.

Theorem 15.1 (necessary condition). Let x^* be a local minimum for (DNLP). Then to every $d \in D(x^*)$ there correspond vectors $\lambda = (\lambda_0, \lambda_1, \ldots, \lambda_m)^T$ and $\mu = (\mu_1, \mu_2, \ldots, \mu_p)^T$, not both zero, such that

$$\lambda_i \geqslant 0, \qquad i = 0, 1, \ldots, m \tag{0.16}$$

$$\nabla_x L(x^*, \lambda, \mu) = 0 \tag{0.17}$$

$$d^T \nabla_x^2 L(x^*, \lambda, \mu) d \geqslant 0 \tag{0.18}$$

$$\lambda_i f^i(x^*) = 0, \qquad i = 0, 1, \ldots, m \tag{0.19}$$

$$\lambda_i \nabla f^i(x^*) d = 0, \qquad i \in I_0(x^*). \tag{0.20}$$

Theorem 16.1 (sufficient condition). A feasible solution x^* is an isolated local minimum if either $D(x^*) = \{0\}$, or to every $0 \neq d \in D(x^*)$ there correspond vectors λ, μ satisfying (0.16), (0.17), (0.19), (0.20), and

$$d^T \nabla_x^2 L(x^*, \lambda, \mu) d > 0. \tag{0.18$'$}$$

These results generalize and extend known optimality conditions. For example, the special case $d = 0$ in Theorem 15.1 yields the Mangasarian-Fromovitz condition (MANGASARIAN-FROMOWITZ [67]). The classical second-order necessary condition of the Kuhn-Tucker type (see, for example, FIACCO and McCORMICK [68]) corresponds to the existence, in Theorem 15.1, of fixed multipliers λ, μ for *all* $d \in D(x^*)$, as guaranteed by certain regularity conditions.

PART 1
CONVEX PROGRAMMING

1
CHARACTERIZING OPTIMAL SOLUTIONS

1 DIRECTIONS

1.1 Definition. Let S be a convex set in R^n and let x^* be in cl (S), the *closure* of S. Then a vector $d \in R^n$ is called a *feasible direction for S at* x^* if there is a $T > 0$ such that

$$x^* + td \in S \qquad \text{for all} \quad 0 < t \leqslant T.$$

We denote the *set of feasible directions for* S *at* x^* by $F(S, x^*)$. In particular, if $f : R^n \to R$ is a convex function, then its *level sets*

$$\{x : f(x) < \alpha\} \qquad \text{and} \qquad \{x : f(x) \leqslant \alpha\}$$

are convex for all real α (see also Ex. 1.1). If x^* is not a minimum point for f, then every feasible direction for $\{x : f(x) < f(x^*)\}$ at x^* is a direction along which the function f decreases, at least initially. Such a direction is called a *direction of descent of f at* x^*.

We now define directions of descent, and associated directions, for general functions.

1.2 Definitions. Let f be a function: $R^n \to \overline{R}$ and let $x^* \in \text{dom } f$. For each of the relations

$$\text{``RELATION''} \underline{\underline{\Delta}} \text{``} < \text{'', ``} > \text{'', ``} = \text{'', etc.,}$$

we denote

$$D_f^{\text{RELATION}}(x^*) = \{d \in R^n : \exists T > 0 \ni f(x + td)\text{RELATION } f(x),$$

$$\forall t \in (0, T)\}. \tag{1.1}$$

13

In particular we call

$$
\left.\begin{matrix}
D_f^{<}(x^*) \\
D_f^{=}(x^*) \\
D_f^{\leqslant}(x^*) \\
D_f^{\geqslant}(x^*)
\end{matrix}\right]
\quad \text{the } \textit{set of directions of} \quad
\left\{\begin{matrix}
\textit{descent} \\
\textit{constancy} \\
\textit{nonascent} \\
\textit{nondescent}
\end{matrix}\right\}
\quad \textit{of } f \textit{ at } x^*.
$$

If $\{f^k : k \in \Omega\}$ is a set of functions indexed by a set Ω, we use the following abbreviations

$$
D_k^{\text{RELATION}}(x^*) \triangleq D_{f^k}^{\text{RELATION}}(x^*)
$$

$$
D_\Omega^{\text{RELATION}}(x^*) \triangleq \bigcap_{k \in \Omega} D_k^{\text{RELATION}}(x^*),
$$

where for $\Omega = \varnothing$, the empty set, $D_\varnothing^{\text{RELATION}}$ is interpreted as R^n.

By definition, the sets $D_f^{=}(x^*)$ and $D_f^{\leqslant}(x^*)$ are *cones* (i.e., sets closed under multiplication by non-negative scalars) and the set $D_f^{<}(x^*)$ is a *blunt cone* (i.e., a set closed under multiplication by positive scalars; $D_f^{<}(x^*)$ does not contain the origin $d = 0$). Little more can be said about these sets, except when convexity is assumed.

1.3 Lemma. Let f be a convex function: $R^n \to R$ and let $x^* \in \text{dom } f$. Then

(a) $\text{comp } D_f^{<}(x^*) = D_f^{\geqslant}(x^*)$; in words: at x^*, the complement of the set of directions of descent is the set of directions of nondescent;

(b) $D_f^{\leqslant}(x^*) = D_f^{<}(x^*) \cup D_f^{=}(x^*)$;

(c) the blunt cone $D_f^{<}(x^*)$ is convex;

(d) the cone $D_f^{\leqslant}(x^*)$ is convex;

(e) $\text{conv } D_f^{=}(x^*) \subset D_f^{\leqslant}(x^*)$

where "conv" denotes "convex hull."

Proof.

(a) Clearly

$$
D_f^{\geqslant}(x^*) \subset \text{comp } D_f^{<}(x^*).
$$

To prove the converse we note that since f is convex,

$$
f(x^* + d) < f(x^*), \qquad 0 < t \leqslant 1 \Rightarrow f(x^* + td) < f(x^*). \tag{1.2}
$$

(This property, and consequently this lemma, holds for the wider class of *strictly quasiconvex functions*, see Ex. 1.2.) Suppose that

$d \notin D_f^{\geq}(x^*)$, that is, for every $T > 0$ there is a $t_1 \in (0, T)$ such that $f(x^* + t_1 d) < f(x^*)$. Then by (1.2),

$$f(x^* + td) < f(x^*) \qquad \text{for all} \quad 0 < t \leq t_1,$$

proving that $d \in D_f^{<}(x^*)$. Therefore

$$\text{comp } D_f^{<}(x^*) \subset D_f^{\geq}(x^*).$$

(b) Clearly both $D_f^{<}(x^*)$ and $D_f^{=}(x^*)$ are contained in $D_f^{\leq}(x^*)$, so that

$$D_f^{<}(x^*) \cup D_f^{=}(x^*) \subset D_f^{\leq}(x^*). \tag{1.3}$$

Since

$$D_f^{=}(x^*) = D_f^{\leq}(x^*) \cap D_f^{\geq}(x^*),$$

it follows from (a) that

$$D_f^{=}(x^*) = D_f^{\leq}(x^*) \cap \text{comp } D_f^{<}(x^*)$$

which, combined with (1.3), proves (b).

(c) Let $d^0, d^1 \in D_f^{<}(x^*)$ and assume that

$$f(x^* + td^i) < f(x^*), \qquad \forall t \in (0, 1], \qquad i = 0, 1. \tag{1.4}$$

For $0 \leq \lambda \leq 1$ denote

$$d^\lambda \underline{\Delta} \lambda d^1 + (1 - \lambda) d^0. \tag{1.5}$$

Suppose $d^\lambda \notin D_f^{<}(x^*)$. Then by (a), $d^\lambda \in D_f^{\geq}(x^*)$, so there is $T_\lambda > 0$ such that

$$f(x^* + td^\lambda) \geq f(x^*), \qquad \forall t \in (0, T_\lambda]. \tag{1.6}$$

Let

$$T = \min\{1, T_\lambda\}.$$

Then by (1.5),

$$x^* + Td^\lambda = \lambda(x^* + Td^1) + (1 - \lambda)(x^* + Td^0); \tag{1.7}$$

but by (1.4) and (1.6),

$$f(x^* + Td^\lambda) > \max\{f(x^* + Td^1), f(x^* + Td^0)\}, \tag{1.8}$$

contradicting the convexity of f.

(d) Let $d^0, d^1 \in D_f^{\leq}(x^*)$ and assume that

$$f(x^* + td^i) \leq f(x^*), \qquad \forall t \in (0, 1], \qquad i = 0, 1.$$

For $0 \leq \lambda \leq 1$ define d^λ by (1.5). If $d^\lambda \notin D_f^{\leq}(x^*)$ then, in particular, there is a $0 < T < 1$ such that

$$f(x^* + Td^\lambda) > f(x^*).$$

For this T we have (1.7), and the contradiction (1.8) as before.

(e) Follows from (b) and (d).   ∎

Lemma 1.3(a) ("d is a direction of nondescent if and only if it is not a direction of descent") may sound redundant, but as Ex. 1.6 shows, it need not hold for general functions.

1.4 Lemma. Let f be a convex function: $R^n \rightarrow R$, and let x^* be a point where f is finite. Then

(a) $D_f^<(x^*) = \{d : f'(x^*,d) < 0\}$; in words: d is a descent direction if and only if the directional derivative

$$f'(x^*,d) \triangleq \lim_{t \rightarrow 0^+} \frac{f(x^* + td) - f(x^*)}{t}$$

is negative;

(b) $D_f^=(x^*) \subset \{d : f'(x^*,d) = 0\}$;

(c) $D_f^\leqq(x^*) = \{d : f'(x^*,d) \gneqq 0 \text{ where } f'(x^*,d) = 0 \text{ only if } d \in D_f^=(x^*)\}$.

Proof.

(a) The inclusion

$$D_f^<(x^*) \supset \{d : f'(x^*,d) < 0\}$$

holds regardless of convexity. To prove the converse, recall that for a convex function (ROCKAFELLAR [70a, Section 23])

$$f'(x^*,d) = \inf_{t>0} \frac{f(x^* + td) - f(x^*)}{t}.$$

Therefore

$$f'(x^*,d) \geq 0 \Rightarrow f(x^* + td) \geq f(x^*), \qquad \forall t > 0$$

proving that

$$D_f^<(x^*) \subset \{d : f'(x^*,d) < 0\}.$$

(b) Obvious.

(c) Follows from (a), (b), and Lemma 1.3(b).   ∎

For a convex function f, the cone $D_f^=(x^*)$ need not be convex (see Ex. 1.7), except if f is differentiable at x^*. Another reason for restating Lemma 1.4 in the differentiable case is that the resulting Corollary 1.5 holds for the wider class of *pseudoconvex functions* (see Ex. 1.4). This is easily shown by proving Corollary 1.5 directly, and not as a consequence of Lemma 1.4.

1.5   Corollary.   Let f be a convex function: $R^n \to R$ and let $x^* \in \operatorname{dom} f$ be a point where f is differentiable. Then

(a) $D_f^<(x^*) = \{d : \nabla f(x^*)^T d < 0\}$;

(b) $D_f^=(x^*)$ is a convex cone and

$$D_f^=(x^*) \subset \{d : \nabla f(x^*)^T d = 0\}; \tag{1.9}$$

(c)

$$D_f^\le(x^*) = \begin{cases} d: & \nabla f(x^*)^T d \le 0, \\ & \nabla f(x^*)^T d = 0 \Rightarrow d \in D_f^=(x^*) \end{cases}.$$

Proof.   Only the convexity of $D_f^=(x^*)$ is new and needs proving. Let $d^0, d^1 \in D_f^=(x^*)$ and assume that

$$f(x^* + td^i) = f(x^*), \qquad \forall t \in (0,1], \qquad i = 0, 1.$$

For $0 \le \lambda \le 1$ let d^λ be defined by (1.5). Then for every $0 \le t \le 1$ it follows from the convexity of f that

$$f(x^* + td^\lambda) \le \lambda f(x^* + td^1) + (1 - \lambda) f(x^* + td^0)$$
$$= f(x^*)$$

and conversely, since f is also differentiable, by the gradient inequality:

$$f(x^* + td^\lambda) \ge f(x^*) + t\nabla f(x^*)^T d^\lambda$$
$$= f(x^*) + t\nabla f(x^*)^T [\lambda d^1 + (1 - \lambda)d^0]$$
$$= f(x^*), \qquad \text{by (1.9)}.$$

Therefore $d^\lambda \in D_f^=(x^*)$, proving the convexity of $D_f^=(x^*)$.   ∎

Examples 1.7 and 1.8 show that little more can be said about the set of constancy directions, $D_f^=(x^*)$, which can become quite unpleasant even if f is convex. This may be the reason that the sets $D_f^=(x^*)$ have been avoided in the literature. To reassure the reader we now show that for a wide class of functions (which includes all analytic convex functions and therefore most functions of practical interest) the sets $D_f^=(x^*)$ are merely subspaces, in fact, the same subspace for all x^*.

We start by noting that if the function $f: R^n \to R$ is *affine*, that is, if

$$f(x) = a^T x + \alpha \tag{1.10}$$

for some $a \in R^n$, $\alpha \in R$, then

$$D_f^=(x^*) = \{d : a^T d = 0\} \quad \text{and} \quad D_f^<(x^*) = \{d : a^T d \leqslant 0\}$$
$$\text{for all} \quad x^* \in R^n. \tag{1.11}$$

A natural extension of the affine case is considered in the following.

1.6   Lemma.   Let the function $f: R^n \to R$ be given as

$$f(x) = h(Ax + b) + a^T x + \alpha \tag{1.12}$$

where

$$A \text{ is an } m \times n \text{ matrix,}$$
$$b \in R^m, \quad a \in R^n, \quad \alpha \in R,$$

and $h: R^m \to R$ is a function whose graph contains no line segment. Then

$$D_f^=(x^*) = N\left(\begin{bmatrix} A \\ a^T \end{bmatrix}\right), \tag{1.13}$$

the null space of the $(m + 1) \times n$ matrix $\begin{bmatrix} A \\ a^T \end{bmatrix}$.

Proof. If $d \in N(\begin{bmatrix} A \\ a^T \end{bmatrix})$, that is, if $Ad = 0$ and $a^T d = 0$, then by (1.12), $d \in D_f^=(x^*)$, regardless of the special assumptions on h. Conversely, let $d \in D_f^=(x^*)$, that is, let $T > 0$ be such that for all $0 < t \leqslant T$

$$f(x^* + td) - f(x^*) = h((Ax^* + b) + tAd) - h(Ax^* + b) + ta^T d$$
$$= 0, \tag{1.14}$$

which shows that h is affine on the interval

$$[Ax^* + b, Ax^* + b + TAd];$$

a contradiction unless

$$Ad = 0,$$

which, by (1.14), implies that

$$a^T d = 0.$$

∎

Lemma 1.6 applies, in particular, to the class of faithfully convex functions. A convex function $f: R^n \to R$ is called *faithfully convex* if whenever f is affine on a line segment, f is affine on the entire line containing the segment. ROCKAFELLAR [70b] showed that a function $f: R^n \to R$ is faithfully convex if and only if it is of the form (1.12), where $h: R^m \to R$ is strictly convex. The class of faithfully convex functions includes all analytic convex functions. In particular, the posynomial functions of geometric programming (DUFFIN, PETERSON, and ZENER [67]) are transformed into faithfully convex functions by the exponential transformation of all variables (see Ex. 1.10).

An algorithm for constructing $D_f^=$ for differentiable faithfully convex functions is outlined in Ex. 1.11.

We now come to the main result of this chapter, relating feasible directions for convex sets (defined by convex inequality constraints) to the nonascent directions of the constraint functions.

1.7 Theorem. Let the convex set S be represented by inequality constraints

$$S = \{ x : f^k(x) \leqslant 0, \quad k \in \mathcal{P} \} \tag{1.15}$$

where $\mathcal{P}$ is a finite index set and $\{ f^k : k \in \mathcal{P} \}$ are convex functions: $R^n \to R$. Let $x^* \in S$ and let

$$\mathcal{P}(x^*) = \{ k \in \mathcal{P} : f^k(x^*) = 0 \} \tag{1.16}$$

denote the *index set of binding constraints at* x^*. Then the set of feasible directions for S at x^*, $F(S, x^*)$, is given by

$$F(S, x^*) = D_{\mathcal{P}(x^*)}^{\leqslant}(x^*), \tag{1.17}$$

the intersection of the nonascent directions for all the binding constraints at x^*.

Proof. Let $d \in F(S, x^*)$, that is, there is a $T > 0$ such that
$$x^* + td \in S, \quad \forall t \in (0, T];$$
in particular, for every $k \in \mathcal{P}(x^*)$,
$$f^k(x^* + td) \leqslant 0 = f^k(x^*) \quad \text{for all} \quad 0 < t \leqslant T,$$
proving that $d \in D_{\mathcal{P}(x^*)}^{\leqslant}(x^*)$.

Conversely, let $d \in D_{\mathcal{P}(x^*)}^{\leqslant}(x^*)$. Then for every $k \in \mathcal{P}(x^*)$, there is a $T_k > 0$ such that

$$f^k(x^* + td) \leqslant f^k(x^*) = 0, \quad \forall t \in (0, T_k]. \tag{1.18}$$

For $l \in \mathcal{P} \setminus \mathcal{P}(x^*)$ [i.e., $f^l(x^*) < 0$] there is, by continuity, a $T_l > 0$ such that

$$f^l(x^* + td) \leqslant 0, \quad \forall t \in (0, T_l]. \tag{1.19}$$

Let

$$T = \min\left\{ \min_{k \in \mathcal{P}(x^*)} \{ T_k \}, \quad \min_{l \in \mathcal{P} \setminus \mathcal{P}(x^*)} \{ T_l \} \right\}.$$

Then, from (1.18) and (1.19), for every $k \in \mathcal{P}$

$$f^k(x^* + td) \leq 0, \qquad \forall t \in (0, T]$$

proving that $d \in F(S, x^*)$.   ∎

How does this result change if $\mathcal{P}$ is not finite? An examination of the proof shows that the part

$$F(S, x^*) \subset D^{\leq}_{\mathcal{P}(x^*)}(x^*)$$

still holds, but that the proof of the converse is no longer valid. Example 1.13 shows that in the case of infinite $\mathcal{P}$,

$$F(S, x^*) \not\supset D^{\leq}_{\mathcal{P}(x^*)}(x^*)$$

is possible.

Exercises and Examples

SOME GENERALIZATIONS OF CONVEXITY

1.1   Quasiconvex functions.   A function $f: R^n \to R$, defined on a convex set in R^n, is called *quasiconvex* if all its level sets

$$\{ x : f(x) \leq \alpha \}, \qquad \alpha \in R$$

are convex. The following are equivalent:

(a) f is quasiconvex;
(b) $f(x + d) \leq f(x), 0 \leq t \leq 1 \Rightarrow f(x + td) \leq f(x) \; \forall x, x + d \in \text{dom } f$;
(c) $f[\lambda x + (1 - \lambda)y] \leq \max\{ f(x), f(y) \}$ for all $0 \leq \lambda \leq 1$; $x, y \in \text{dom } f$.

1.2   Strictly quasiconvex functions.   A function $f: R^n \to R$, defined on a convex set in R^n, is called *strictly quasiconvex* if for all $x, x + d \in \text{dom } f$,

$$f(x + d) < f(x), \qquad 0 < t \leq 1 \Rightarrow f(x + td) < f(x).$$

1.3   If f is strictly quasiconvex and x^* is a local minimum of f, then x^* is a global minimum of f.

Proof.   Let x^* be a local minimum of f and let y be a point at which

$f(y) < f(x^*)$. Then for all $0 < \lambda < 1$,

$$f(\lambda x^* + (1 - \lambda)y) < f(x^*),$$

contradicting the local minimality of x^*. ■

1.4 Pseudoconvex functions. A function $f: R^n \to R$, defined and differentiable on a convex set Γ in R^n, is called *pseudoconvex* on Γ if for all $x, y \in \Gamma$

$$\nabla f(x)^T(y - x) \geqslant 0 \Rightarrow f(y) \geqslant f(x).$$

If f is a pseudoconvex function, then, by definition, a sufficient condition for x^* to be a global minimum in Γ is

$$\nabla f(x^*)^T d \geqslant 0 \qquad \forall x^* + d \in \Gamma.$$

1.5 Some relations and examples for the above kinds of convexity. The following implications hold for functions: $R^n \to R$.

  (a) Convexity $\Rightarrow$ quasiconvexity, strict quasiconvexity.
  (b) Convexity and differentiability $\Rightarrow$ pseudoconvexity.
  (c) Pseudoconvexity $\Rightarrow$ strict quasiconvexity.

The example $f(x) = x^3$ shows that a differentiable strictly quasiconvex function need not be pseudoconvex.

  (d) Strict quasiconvexity and lower semicontinuity $\Rightarrow$ quasiconvexity.

The example

$$f(x) = \begin{cases} 1, & x = 0 \\ 0, & x \neq 0 \end{cases}$$

shows that lower semicontinuity is essential here.

For further reading on these generalized convex function and others see MANGASARIAN [69, Chapter 9], MARTOS [75, Chapters 3 and 7], AVRIEL [76, Chapter 6], PONSTEIN [67], KARAMARDIAN [67], and GREENBERG and PIERSKALLA [71].

1.6 Convexity is required in Lemma 1.3(a). Let $f: R \to R$ be given by

$$f(x) = \begin{cases} x \sin \dfrac{1}{x}, & x \neq 0 \\ 0, & x = 0. \end{cases}$$

Then at $x^* = 0$,

$$D_f^<(x^*) = \varnothing, \qquad D_f^>(x^*) = \{0\}$$

showing that, in general,

$$\text{comp } D_f^<(x^*) \neq D_f^>(x^*).$$

1.7 Directions of constancy. If f is convex, but f is not differentiable at x^*, the cone $D_f^=(x^*)$ need not be convex. Let $f : R^2 \to R$ be the l_1-norm

$$f(x_1, x_2) = |x_1| + |x_2|.$$

Then at $x^* = (1, 0)$ the set $D_f^=(x^*)$ is the union of the two halflines

$$x_2 = x_1, \qquad x_1 \leqslant 0$$

and

$$x_2 = -x_1, \quad x_1 \leqslant 0.$$

1.8 An example of the nonclosedness of the cones $D_f^=$. Let $f : R^2 \to R$ be

$$f(x_1, x_2) = \begin{cases} \left(x_1^2 + x_2^2 - 1\right)^2, & \text{if } x_1^2 + x_2^2 \geqslant 1 \\ 0, & \text{otherwise,} \end{cases}$$

and let x^* be on the circle $x_1^2 + x_2^2 = 1$. Then

$$D_f^=(x^*) = \{(d_1, d_2) : d_1 x_1^* + d_2 x_2^* < 0\} \cup \{(0, 0)\}.$$

EXAMPLES OF FAITHFULLY CONVEX FUNCTIONS

1.9 Strictly convex restrictions. For a function $f^k : R^n \to R$ we denote by $[k]$ the index set of the variables on which f^k actually depends, that is, $j \in [k]$ if there are $n - 1$ values

$$\xi_1, \xi_2, \ldots, \xi_{j-1}, \xi_{j+1}, \ldots, \xi_n$$

such that the function

$$f^k(\xi_1, \xi_2, \ldots, \xi_{j-1}, \cdot, \xi_{j+1}, \ldots, \xi_n)$$

is not constant. For a vector $x = [x_j] \in R^n$ we then denote by $x_{[k]}$ the subvector

$$x_{[k]} = [x_j], \qquad j \in [k].$$

The *restriction* of f is the function

$$f^{[k]} : R^{\text{card}[k]} \to R$$

defined by $f^{[k]}(x_{[k]}) = f^k(x) \qquad \forall x \in R^n.$

The function $f^k: R^n \to R$ is said to have a *strictly convex restriction* if its restriction $f^{[k]}$ is a strictly convex function of its variables $x_{[k]}$. Then by a straightforward application of Lemma 1.6, for all $x^* \in \text{dom } f^k$,

$$D_{f^k}^=(x^*) = \{d \in R^n : d_{[k]} = 0\}.$$

For example, let $f^3: R^5 \to R$ be given by

$$f^3(x_1, x_2, x_3, x_4, x_5) = e^{-x_2} + (x_5)^2.$$

Then

$$[3] = \{2, 5\}$$
$$x_{[3]} = (x_2, x_5),$$

and the restriction

$$f^{[3]}(x_2, x_5) = e^{-x_2} + (x_5)^2$$

is strictly convex (although f^3 itself is not), and for all $x^* \in R^5$,

$$D_{f^3}^=(x^*) = \{(d_1, 0, d_3, d_4, 0)^T : d_1, d_3, d_4 \in R\}.$$

1.10 Posynomials. A function $f: R^n \to R$ is called a *posynomial* if f is of the form

$$f(x_1, \ldots, x_n) = \sum_{i=1}^m c_i \prod_{j=1}^n x_j^{a_{ij}} \tag{1.20}$$

where the coefficients c_i are positive and the exponents a_{ij} are real. By using the exponential transformation

$$x_j = e^{y_j}, \quad j = 1, \ldots, n$$

the posynomial (1.20) is transformed into the faithfully convex function

$$\hat{f}(y_1, \ldots, y_n) = \sum_{i=1}^n c_i \exp \sum_{j=1}^n a_{ij} y_j$$

whose directions of constancy, at any point y^*, are by Lemma 1.6

$$D_{\hat{f}}^=(y^*) = \{d : Ad = 0\}$$

where A is the $m \times n$ matrix $[a_{ij}]$.

Calculating the Cone of Directions of Constancy. When $f: R^n \to R$ is a differentiable faithfully convex function, but its explicit representation 1.12 is not known, one can construct $D_f^=$ in at most n iterations by the following algorithm. For more details and proofs see WOLKOWICZ [78].

1.11 Algorithm. Let E^s denote the unit basis in R^s, that is, the collection of the s unit vectors of R^s.

Initialization.   Denote $A_0 = I_{n \times n}$ and $f^1 = f$.

Iteration k $(1 \leqslant k \leqslant n)$.   Find $x^k \in \{0\} \cup E^{n-k+1}$ such that $\nabla f^k(x^k) \neq 0$.

Case I   A suitable x^k has been found and $k < n$. Construct $A_k \in R^{(n-k+1) \times (n-k)}$ such that $R(A_k) = N[\nabla f^k(x^k)^T]$ and $f^{k+1} = f^k \circ A_k$. (Here $R(A_k)$ denotes the range space of A_k.) Set $k_{NEW} = k + 1$ and repeat.

Case II   A suitable x^k has been found and $k = n$. Stop; $D_f^= = \{0\}$.

Case III   For every $x \in \{0\} \cup E^{n-k+1}$, $\nabla f^k(x) = 0$. Stop; $D_f^=$ $= R(A_0 A_1 \ldots A_{k-1})$, that is, the range space of the product of k matrices.

1.12   Example for Algorithm 1.11.   Consider

$$f(x) = -\left[1 + (x_1 + x_2)^2\right]^{1/2} + x_1 + x_2.$$

Initialization

$$A_0 = \begin{bmatrix} 1 & 0 \\ 0 & 1 \end{bmatrix}, \qquad f^1 = f.$$

Iteration $k = 1$.   Choose

$$x^1 = \begin{bmatrix} 0 \\ 0 \end{bmatrix} \in \{0\} \cup E^2 = \left\{ \begin{bmatrix} 0 \\ 0 \end{bmatrix}, \begin{bmatrix} 1 \\ 0 \end{bmatrix}, \begin{bmatrix} 0 \\ 1 \end{bmatrix} \right\}$$

and observe that $\nabla f^1(x^1) = (1, 1)^T \neq 0$.

Since $k < n$, we have Case I. Now

$$N\left(\nabla f^1(x^1)^T\right) = \left\{ \begin{bmatrix} \alpha \\ -\alpha \end{bmatrix} : \alpha \in R \right\};$$

hence

$$A_1 = \begin{bmatrix} 1 \\ -1 \end{bmatrix} \quad \text{and} \quad f^2(x) = f^1(A_1 x) = -1.$$

Iteration $k = 2$.   Since $\nabla f^2(x) = 0$ for every $x \in \{0\} \cup E^1 = \{0, 1\}$, we have Case III and conclude that

$$D_f^= = R(A_0 A_1) = \left\{ \begin{bmatrix} d_1 \\ -d_1 \end{bmatrix} : d_1 \in R \right\}.$$

1.13   Infinitely many constraints.   The following example shows that Theorem 1.7 may fail if the number of constraints is infinite.

Let S in R^2 be given by

$$S = \left\{ (x_1, x_2) : \begin{array}{l} f(x_1, x_2, t) \leqslant 0 \\ g(x_1, x_2, t) \leqslant 0 \end{array}, \quad t \in [0, 1] \right\}$$

where

$$f(x_1, x_2, t) = x_1 + x_2 t - t^2$$

$$g(x_1, x_2, t) = -x_1 - x_2 t - t.$$

Then

$$S = \{ (0, x_2) : -1 \leqslant x_2 \leqslant 0 \}$$

and at the point $x^* = (0, 0)$, the binding constraints are

$$f(x_1, x_2, 0) = x_1 \leqslant 0$$

$$g(x_1, x_2, 0) = -x_1 \leqslant 0$$

for which both vectors

$$d^1 = (0, 1)^T, \qquad d^2 = (0, -1)^T$$

are nonascent directions, but only d^2 is a feasible direction.

2   DUAL SETS AND SEPARATION

This section contains selected results from convex analysis that are used later to derive optimality conditions in a dual form.

2.1   Definition.   Let S be a nonempty set in R^n. The *dual set* (also called "polar set") of S, denoted by S^*, is the set of vectors $y \in R^n$ forming non-negative inner products with every $x \in S$,

$$S^* = \{ y \in R^n : x \in S \Rightarrow y^T x \geqslant 0 \}.$$

Some properties of dual sets are listed below without proof (see, for example, BEN-ISRAEL [69]).

2.2   Lemma.   Let S be a nonempty subset of R^n. Then

(a) S^* is a closed convex cone;

(b) $S^* = (\text{conv } S)^*$
$= (\text{conh } S)^*$, where "conh" is conical hull
$= (\text{cl } S)^*$, where "cl" is closure;

(c)   if S is an open set, and if $0 \neq y \in S^*$, then

$$x \in S \Rightarrow y^T x > 0;$$

(d) if S is the halfspace

$$S = \{x : a^T x \leqslant 0\} \qquad \left(\text{or } S = \{x : a^T x < 0\}\right)$$

then S^* is the halfline

$$S^* = \{\lambda a : \lambda \leqslant 0\};$$

(e) if S is a subspace of R^n, then $S^* = S^\perp$, the orthogonal complement of S;

(f) if $S \subset T$, then $S^* \supset T^*$;

(g) if S and T are convex sets, then $(S \cap T)^* = \text{cl}(S^* + T^*)$. Moreover, if S and T are polyhedral then "closure" can be omitted. ∎

The separation theorem most suitable for our purposes is the following variant of the Dubovitskii-Milyutin theorem (DUBOVITSKII and MILYUTIN [63], see also HOLMES [75, p. 116] and GIRSANOV [72]). ∎

2.3 Theorem (Dubovitskii-Milyutin). Let $S_0, S_1, \ldots, S_l$ be convex sets with

$$0 \in \text{cl } S_k, \qquad k = 0, 1, \ldots, l$$

and S_k, $k = 1, 2, \ldots, l$ are open. Then the intersection

$$\bigcap_{k=0}^{l} S_k = \varnothing$$

is empty if and only if there are vectors

$$y^k \in S_k^*, \qquad k = 0, 1, \ldots, l$$

not all zero, such that

$$\sum_{k=0}^{l} y^k = 0.$$

∎

A consequence of Theorem 2.3 is the following *theorem of the alternatives*.

2.4 Corollary. Let A be an $m \times n$ matrix, c an n-vector, and S a closed convex cone in R^n. Then the system

$$c^T x < 0, \qquad Ax = 0, \qquad x \in S \tag{2.1}$$

has no solution if and only if

$$c \in \text{cl}\left[R(A^T) + S^* \right] \tag{2.2}$$

where $R(A^T)$ is the range space of A^T. If S is polyhedral, the "cl" can be omitted.

Proof. Denote

$$S_0 = \{x : Ax = 0, \qquad x \in S\}$$
$$= N(A) \cap S$$

and

$$S_1 = \{x : c^T x < 0\}.$$

The inconsistency of system (2.1) is just $S_0 \cap S_1 = \varnothing$. Both sets are convex cones and S_0 is open, so Theorem 2.3 applies, that is, there exist vectors

$$0 \neq y^0 \in S_0^*, \qquad 0 \neq y^1 \in S_1^* \qquad \text{such that} \quad y^0 + y^1 = 0. \qquad (2.3)$$

Now

$$S_0^* = \text{cl}\big[R(A^T) + S^*\big], \quad \text{by Lemma 2.2}(e) \text{ and } 2.2(g)$$
$$S_1^* = \{\lambda c : \lambda \leqslant 0\}, \qquad \text{by Lemma 2.2}(d);$$

hence the consistency of (2.3) is equivalent to (2.2). If S is polyhedral, so is S^*. This implies that $R(A^T) + S^*$ is also polyhedral and hence closed. ∎

2.5 Definition. A vector $y \in R^n$ is a *subgradient* of the function $f: R^n \to R$ at x^* if

$$f(z) \geqslant f(x^*) + y^T(z - x^*), \qquad \forall z.$$

The set of all subgradients of f at x^* is denoted by $\partial f(x^*)$. If f is convex and differentiable then $\partial f(x^*) = \{\nabla f(x^*)\}$.

2.6 Lemma (ROCKAFELLAR [70a, Chapter 23]). Let $f: R^n \to R$ be a convex function and let x^* be a point in the relative interior of dom f such that $f(x^*)$ is not the minimum of f. Then

(a) $(D_f^{\leqslant}(x^*))^* = \text{cl}\{\mu \partial f(x^*) : \mu \leqslant 0\}$;

(b) if, further, x^* is in the interior of dom f, then

$$(D_f^{\leqslant}(x^*))^* = \{\mu \partial f(x^*) : \mu \leqslant 0\}.$$

∎

3 CHARACTERIZATIONS OF OPTIMALITY

Consider the convex program (P)

$$\min f^0(x)$$

s.t.

$$f^k(x) \leqslant 0, \qquad k \in \mathscr{P} \underset{=}{\Delta} \{1, 2, \ldots, p\}$$

where $f^k : R^n \to R$, $k = 0, 1, \ldots, p$ are continuous convex functions. The *feasible set* of (P) is

$$S = \left\{ x : f^k(x) \leqslant 0, \qquad k \in \mathcal{P} \right\}$$

and the *index set of binding constraints at* $x \in S$ is

$$\mathcal{P}(x) = \left\{ k \in \mathcal{P} : f^k(x) = 0 \right\}.$$

In convex programming, optimality is characterized by the absence of feasible directions that improve the objective function (*usable directions*).

3.1 Lemma. A feasible point x^* of program (P) is optimal if and only if

$$D_0^{\leqslant}(x^*) \cap F(S, x^*) = \varnothing \tag{3.1}$$

or, more explicitly,

$$D_0^{\leqslant}(x^*) \cap D_{\mathcal{P}(x^*)}^{\leqslant}(x^*) = \varnothing. \tag{3.2}$$

Proof. If. Suppose x^* is not optimal, that is, there is a nonzero vector d such that

$$x^* + d \in S \tag{3.3}$$

and

$$f^0(x^* + d) < f^0(x^*). \tag{3.4}$$

Since $x^* \in S$ it follows from (3.3) and the convexity of S that

$$x^* + \alpha d \in S \qquad \forall \quad 0 < \alpha \leqslant 1,$$

so that $d \in F(S, x^*)$. Similarly, from (3.4) and the convexity of f^0 it follows that

$$f^0(x^* + \alpha d) < f^0(x^*) \qquad \forall \quad 0 < \alpha \leqslant 1,$$

and consequently $d \in D_{\mathcal{P}}^{\leqslant}(x^*)$, proving that

$$D_0^{\leqslant}(x^*) \cap F(S, x^*) \neq \varnothing.$$

Only if. Obvious. This proves (3.1). Relation (3.2) follows by (1.17). ∎

The first characterization theorem follows.

3.2 THEOREM. A feasible solution x^* of (P) is optimal if and only if for every subset Ω of $\mathcal{P}(x^*)$, the following intersection is empty

$$D(\Omega, x^*) \triangleq D_0^{\leqslant}(x^*) \cap D_{\mathcal{P}(x^*)\setminus\Omega}^{\leqslant}(x^*) \cap D_{\Omega}^{=}(x^*) = \varnothing; \tag{3.5}$$

or, dually, for every subset Ω of $\mathcal{P}(x^*)$ there exist vectors

$$y^k \in (D_k^{\leqq}(x^*))^*, \quad k \in \{0\} \cup (\mathcal{P}(x^*)\backslash\Omega) \text{ not all zero such that} \quad (3.6)$$

$$\sum_{k \in \{0\} \cup (\mathcal{P}(x^*)\backslash\Omega)} y^k \in -(D_\Omega^=(x^*))^*. \tag{3.7}$$

Proof. Note that

$$F(S, x^*) = D_{\mathcal{P}(x^*)}^{\leqq}(x^*) \triangleq \bigcap_{k \in \mathcal{P}(x^*)} D_k^{\leqq}(x^*)$$

$$= \bigcap_{k \in \mathcal{P}(x^*)} (D_k^{\leqq}(x^*) \cup D_k^=(x^*)), \quad \text{by Lemma 1.3(b)}$$

$$= \bigcup_{\Omega \subset \mathcal{P}(x^*)} \{D_{\mathcal{P}(x^*)\backslash\Omega}^{\leqq}(x^*) \cap D_\Omega^=(x^*)\}. \tag{3.8}$$

Hence (3.2) is equivalent to

$$D(\Omega, x^*) = \varnothing \quad \text{for every} \quad \Omega \subset \mathcal{P}(x^*),$$

which proves the first part. To prove the second part we note that the term $D_\Omega^=(x^*)$ in (3.5) can be replaced by its convex hull $\text{conv}\, D_\Omega^=(x^*)$. This follows since

$$F(S, x^*) \subset \bigcap_{k \in \mathcal{P}(x^*)} (D_k^{\leqq}(x^*) \cup \text{conv}\, D_k^=(x^*)), \quad \text{by (3.8)}$$

$$\subset \bigcap_{k \in \mathcal{P}(x^*)} (D_k^{\leqq}(x^*) \cup D_k^{\leqq}(x^*)), \quad \text{by Lemma 1.3(e)}$$

$$= \bigcap_{k \in \mathcal{P}(x^*)} D_k^{\leqq}(x^*) = F(S, x^*).$$

Thus (3.5) can be written as

$$\left(\bigcap_{k \in \{0\} \cup (\mathcal{P}(x^*)\backslash\Omega)} S_k \right) \cap S_\Omega = \varnothing \quad \forall \Omega \subset \mathcal{P}(x^*) \tag{3.9}$$

where

$$S_k = D_k^{\leqq}(x^*), \quad k \in \{0\} \cup (\mathcal{P}(x^*)\backslash\Omega) \quad \text{(an open cone)}$$

$$S_\Omega = \text{conv}\, D_\Omega^=(x^*).$$

An application of separation Theorem 2.3 to (3.7) yields the existence of y^ks satisfying the dual condition (3.6), (3.7). ∎

Theorem 3.2 provides 2^{p^*} [p^* is the cardinality of $\mathcal{P}(x^*)$] necessary conditions—one for each Ω. For the particular choice $\Omega = \varnothing$ the following is obtained:

$$\begin{cases} \exists y^k \in (D_k^\le(x^*))^*, \quad k \in \{0\} \cup \mathcal{P}(x^*) \qquad \text{not all zero such that} \\ \qquad\qquad \sum_{k \in \{0\} \cup \mathcal{P}(x^*)} y^k = 0. \end{cases}$$

(3.10)

This is the so-called *Euler-Lagrange equation* in the Dubovitskii-Milyutin formalism (e.g., GIRSANOV [72]), which in the differentiable case reduces to the *Fritz John condition* (e.g., MANGASARIAN [69]).

Under certain conditions (*reduction conditions*) (3.10) is also sufficient for optimality. These are discussed in the Exercises and Examples Section. Here we consider one such condition, namely, *strict convexity* of the constraints.

3.3   Corollary. Let x^* be a feasible point of (P). Assume that the constraint functions, which are binding at x^*, are strictly convex. Then x^* is optimal if and only if (3.10) holds.

Proof. When f^k is strictly convex, then $D_k^=(x^*) = \{0\}$. So $(D_k^=(x^*))^* = R^n$. Therefore for every *nonempty* subset $\Omega \subset \mathcal{P}(x^*)$, (3.6) and (3.7) are trivially satisfied and only the case $\Omega = \emptyset$ remains.   ∎

In general, Theorem 3.2 calls for checking 2^{p^*} systems (one for each Ω). However, as is shown below, only a particular *single* subset has to be checked. This subset, which carries all the information needed to check optimality, is the same for all feasible points and is thus a (global) characteristic of the convex program (P). It is denoted by $\mathcal{P}^=$ and defined as follows.

3.4   Definition. *The minimal index set $\mathcal{P}^=$ of binding constraints* for the feasible set

$$S = \left\{ x : f^k(x) \le 0, \quad k \in \mathcal{P} \right\}$$

is defined as

$$\mathcal{P}^= \underset{=}{\Delta} \left\{ k \in \mathcal{P} : x \in S \Rightarrow f^k(x) = 0 \right\}$$

or, equivalently,

$$\mathcal{P}^= = \bigcap_{x \in S} \mathcal{P}(x).$$

∎

Since $\mathcal{P}^= \subset \mathcal{P}(x)$ for every $x \in S$, we denote the *complement of $\mathcal{P}^=$ relative to $\mathcal{P}(x)$ by*

$$\mathcal{P}^<(x) \triangleq \mathcal{P}(x) \backslash \mathcal{P}^= = \left\{ k \in \mathcal{P}(x) : \exists x^k \in S \ni f^k(x^k) < 0 \right\}.$$

The *complement of $\mathcal{P}^=$ relative to $\mathcal{P}$* is

$$\mathcal{P}^< \triangleq \mathcal{P} \backslash \mathcal{P}^= = \bigcup_{x \in S} \mathcal{P}^<(x) = \left\{ k \in \mathcal{P} : \exists x^k \in S \ni f^k(x^k) < 0 \right\}.$$

Some properties of these sets are collected in the following lemma.

3.5 Lemma. Let x be any point in S. Then

(a) $\mathcal{P}^=$ is the maximal subset Ω of $\mathcal{P}(x)$ with the property that, for every $k \in \Omega$,

$$d \in F(S,x) \Rightarrow d \in D_k^=(x); \tag{3.11}$$

(b) for every subset Ω of $\mathcal{P}^=$

$$D_{\mathcal{P}(x) \backslash \Omega}^{\leqq}(x) \cap D_\Omega^=(x) = D_{\mathcal{P}(x) \backslash \Omega}^{\leqq}(x) \cap \operatorname{conv} D_\Omega^=(x);$$

(c) if $\mathcal{P}^<(x) \neq \varnothing$, then there exists a point $\hat{x} \in S$ such that

$$f^k(\hat{x}) < 0, \qquad k \in \mathcal{P}^<; \tag{3.12}$$

(d) if $\mathcal{P}^<(x) \neq \varnothing$, then for every subset Ω of $\mathcal{P}^=$,

$$D_{\mathcal{P}(x) \backslash \Omega}^{\leqq}(x) \cap D_\Omega^=(x) \begin{cases} \neq \varnothing & \text{if } \Omega = \mathcal{P}^= \\ = \varnothing & \text{if } \Omega \neq \mathcal{P}^=; \end{cases}$$

(e) the cone $D_{\mathcal{P}^=}^=(x)$ is convex (BORWEIN and WOLKOWICZ [79]).

Proof.

(a) First we show that $\Omega = \mathcal{P}^=$ has property (3.11). Indeed, if for some $x \in S, k \in \mathcal{P}^=$

$$d \in F(S,x) \quad \text{and} \quad d \notin D_k^=(x), \quad \text{then } d \in D_k^<(x)$$

and so for a sufficiently small $\alpha > 0$

$$x + \alpha d \in S \quad \text{and} \quad f^k(x + \alpha d) < 0,$$

contradicting $k \in \mathcal{P}^=$. To show the maximality property of $\Omega = \mathcal{P}^=$, let $x \in S$ and $k \in \mathcal{P}(x)$ have property (3.11). If $k \notin \mathcal{P}^=$ then there exists an $\bar{x} \in S$ with $f^k(\bar{x}) < 0$. Therefore the direction $\bar{d} = \bar{x} - x$ is feasible for S at x and a direction of descent for f^k, contradicting (3.11).

(b) Follows from the chain of inclusions

$$D^{\leq}_{\mathscr{P}(x)\backslash\Omega}(x) \cap \operatorname{conv} D^{=}_{\Omega}(x)$$
$$\subset D^{\leq}_{\mathscr{P}(x)\backslash\Omega}(x) \cap D^{\leq}_{\Omega}(x), \qquad \text{by Lemma 1.3(e)}$$
$$\subset D^{\leq}_{\mathscr{P}(x)\backslash\Omega}(x) \cap D^{=}_{\Omega}(x), \qquad \text{by part (a)}$$
$$\subset D^{\leq}_{\mathscr{P}(x)\backslash\Omega}(x) \cap \operatorname{conv} D^{=}_{\Omega}(x).$$

(c) By definition of $\mathscr{P}^{<}(x)$, for every $k \in \mathscr{P}(x)$ there is an $x^k \in S$ such that $f^k(x^k) < 0$. By the convexity of the f^k's, the "center of gravity" of $\{x^k : k \in \mathscr{P}^{<}(x)\}$

$$\hat{x} \triangleq \frac{1}{\operatorname{card} \mathscr{P}^{<}(x)} \sum_{k \in \mathscr{P}^{<}(x)} x^k$$

is feasible and $f^k(\hat{x}) < 0$, $\forall k \in \mathscr{P}^{<}(x)$. The proof now follows from the fact that $\mathscr{P}^{<} = \bigcup_{x \in S} \mathscr{P}^{<}(x)$.

(d) The statement is obvious if $\mathscr{P}^{=} = \varnothing$. Thus suppose that $\mathscr{P}^{=} \neq \varnothing$. Let $\hat{x} \in S$ be the vector mentioned in part (c). Then for every $x \in S$, the direction $\hat{d} = \hat{x} - x$ is in

$$D^{\leq}_{\mathscr{P}(x)\backslash\mathscr{P}^{=}}(x) \cap D^{=}_{\mathscr{P}^{=}}(x)$$

which is thus nonempty, proving the first case. On the other hand, if Ω is a proper subset of $\mathscr{P}^{=}$, then

$$D^{\leq}_{\mathscr{P}(x)\backslash\Omega}(x) \cap D^{=}_{\Omega}(x) \subset D^{\leq}_{\mathscr{P}(x)}(x)$$
$$= D^{\leq}_{\mathscr{P}(x)\backslash\mathscr{P}^{=}}(x) \cap D^{=}_{\mathscr{P}^{=}}(x), \qquad \text{by part (a).}$$

Therefore, unless $D^{\leq}_{\mathscr{P}(x)\backslash\Omega}(x) \cap D^{=}_{\Omega}(x)$ is empty, there is a $k \in \mathscr{P}^{=} \cap [\mathscr{P}(x)\backslash\Omega]$ (nonempty since $\Omega \neq \mathscr{P}^{=}$) for which

$$D^{\leq}_{k}(x) \cap D^{=}_{k}(x) \neq \varnothing,$$

a contradiction.

(e) We show below that $D^{=}_{\mathscr{P}^{=}}(x) = D^{\leq}_{\mathscr{P}^{=}}(x)$. Since the latter set is convex (by Lemma 1.3(d)) this establishes the convexity of $D^{=}_{\mathscr{P}^{=}}(x)$. The inclusion

$$D^{=}_{\mathscr{P}^{=}}(x) \subset D^{\leq}_{\mathscr{P}^{=}}(x)$$

is of course trivial. To prove the reversed inclusion pick $d \in D^{\leq}_{\mathscr{P}^{=}}(x)$. By definition, there exists $\bar{\alpha} > 0$ such that for $x(\alpha) = x + \alpha d$ we have

$$f^k(x(\alpha)) \leq f^k(x) = 0, \qquad \forall k \in \mathscr{P}^{=}, \qquad \forall \alpha \in [0, \bar{\alpha}].$$

From part (c) we know that there exists $\hat{x} \in S$ such that

$$f^k(\hat{x}) < 0, \qquad \forall k \in \mathscr{P}^{<}.$$

These facts and the convexity of the f^k's imply that, for $t > 0$ sufficiently small,

$$f^k(tx(\alpha) + (1 - t)\hat{x}) \leqslant 0, \qquad \forall k \in \mathcal{P}, \qquad \alpha \in [0, \bar{\alpha}]$$

that is,

$$tx(\alpha) + (1 - t)\hat{x} \in S, \qquad t \text{ small}, \qquad \alpha \in [0, \bar{\alpha}].$$

Hence, by definition of $\mathcal{P}^=$, for every $k \in \mathcal{P}^=$:

$$0 = f^k(tx(\alpha) + (1 - t)\hat{x}),$$

$$\leqslant tf^k(x(\alpha)) + (1 - t)f^k(\hat{x}), \qquad \text{by convexity}$$

$$\leqslant tf^k(x(\alpha)), \qquad \text{since } \hat{x} \in S$$

$$\leqslant 0$$

showing that

$$0 = f^k(x(\alpha)) = f^k(x + \alpha d), \qquad \forall \alpha \in [0, \bar{\alpha}].$$

Thus

$$d \in D_{\mathcal{P}^=}^=(x).$$

∎

A characterization of optimality (primal or dual) by a single system, due to Abrams (see ABRAMS and KERZNER [78]) follows.

3.6 Theorem. A feasible solution x^* of (P) is optimal if and only if the following intersection is empty:

$$D(\mathcal{P}^=, x^*) \underline{\triangle} D_0^\leqslant(x^*) \cap D_{\mathcal{P}^<(x^*)}^\leqslant(x^*) \cap D_{\mathcal{P}^=}^=(x^*) = \emptyset;$$

or, dually, there exist vectors

$$0 \neq y^0 \in (D_0^\leqslant(x^*))^*$$

$$y^k \in (D_k^\leqslant(x^*))^*, \qquad k \in \mathcal{P}^<(x^*) \tag{3.13}$$

such that

$$y^0 + \sum_{k \in \mathcal{P}^<(x^*)} y^k \in -(D_{\mathcal{P}^=}^=(x^*))^*.$$

Proof. Only if. Follows from Theorem 3.2 (the case $\Omega = \mathcal{P}^=$).
If. Suppose x^* is not optimal. Then by Lemma 3.1

$$D_0^\leqslant(x^*) \cap D_{\mathcal{P}(x^*)}^\leqslant(x^*) \neq \emptyset.$$

Then let

$$d \in D_0^<(x^*) \cap D_{\mathcal{P}^<(x^*)}^\leqslant(x^*) \cap D_{\mathcal{P}^-}^\leqslant(x^*).$$

Let $\hat{x}$ be a vector satisfying (3.12) and denote $\hat{d} = \hat{x} - x^*$. Then

$$\hat{d} \in D_{\mathcal{P}^<(x^*)}^\leqslant(x^*) \cap D_{\mathcal{P}^-}^=(x^*).$$

For any $0 < \lambda \leqslant 1$ denote $d_\lambda = \lambda\hat{d} + (1-\lambda)d$. Then

$$d_\lambda \in D_{\mathcal{P}^<(x^*)}^\leqslant(x^*) \cap D_{\mathcal{P}^-}^\leqslant(x^*), \qquad \text{by the choice of } d \text{ and } \hat{d}$$

$$= D_{\mathcal{P}^<(x^*)}^\leqslant(x^*) \cap D_{\mathcal{P}^-}^=(x^*), \qquad \text{by Lemma 3.5(a)}$$

and

$$f^0(x^* + \alpha d_\lambda) \leqslant \lambda f^0(x^* + \alpha\hat{d}) + (1-\lambda)f^0(x^* + \alpha d)$$

$$< f^0(x^*) \qquad \text{for sufficiently small } \alpha > 0 \text{ and } \lambda.$$

Therefore

$$d_\lambda \in D_0^<(x^*) \cap D_{\mathcal{P}^<(x^*)}^\leqslant(x^*) \cap D_{\mathcal{P}^-}^=(x^*),$$

contradicting $D(\mathcal{P}^=, x^*) = \emptyset$. We have proved the primal part. To prove the dual part note that the set $D(\mathcal{P}^=, x^*)$ is an intersection of the open convex cones $D_k^<(x^*)$, $k \in \{0\} \cup \mathcal{P}^<(x^*)$ and the convex (see Lemma 3.5(e)) cone $D_{\mathcal{P}^-}^=(x^*)$. Therefore, by the separation Theorem 2.3,

$$D(\mathcal{P}^=, x^*) = \emptyset$$

if and only if

$$\begin{cases} \text{there exist vectors} \quad y^k \in (D_k^<(x^*))^*, \quad k \in \{0\} \cup \mathcal{P}^<(x^*) \text{ not all zero} \\ \text{such that} \\ \displaystyle\sum_{k \in \{0\} \cup \mathcal{P}^<(x^*)} y^k \quad \in -(D_{\mathcal{P}^-}^=(x^*))^*. \end{cases} \tag{3.14}$$

Observe that (3.13) implies (3.14), which proves sufficiency. To prove necessity assume that $y^0 = 0$ in (3.14) and distinguish two cases. If $\mathcal{P}^<(x^*) = \emptyset$, (3.13) is contradicted. If $\mathcal{P}^<(x^*) \neq \emptyset$, then at least one y^k, $k \in \mathcal{P}^<(x^*)$ is nonzero, which, again by Theorem 2.3, implies

$$D_{\mathcal{P}^<(x^*)}^\leqslant(x^*) \cap D_{\mathcal{P}^-}^=(x^*) = \emptyset,$$

contradicting Lemma 3.5(d).   ∎

The dual optimality condition in Theorems 3.2 and 3.6 can be expressed more explicitly in terms of subgradients (gradients in the differentiable case). This is demonstrated here for Theorem 3.6.

3.7 Corollary. Let x^* be a feasible solution of (P) such that

$$x^* \in \bigcap_{k \in \{0\} \cup \mathscr{P}^<(x^*)} \operatorname{int} \operatorname{dom} f^k.$$

Then x^* is optimal if and only if

$$\begin{cases} \text{there exists a vector } y \in \left(D_{\overline{\mathscr{P}}^=}(x^*) \right)^* \\ \text{and nonnegative scalars } \lambda_k \geq 0, k \in \mathscr{P}^<(x^*) \\ \text{such that} \\ \quad y \in \partial f^0(x^*) + \sum_{k \in \mathscr{P}^<(x^*)} \lambda_k \partial f^k(x^*). \end{cases} \tag{3.15}$$

Proof. *If.* Let

$$y = y^0 + \sum_{k \in \mathscr{P}^<(x^*)} \lambda_k y^k \in \left(D_{\overline{\mathscr{P}}^=}(x^*) \right)^* \tag{3.16}$$

where $y^k \in \partial f^k(x^*)$, that is,

$$f^k(x) \geq f^k(x^*) + (x - x^*)^T y^k, \qquad \forall x, \qquad k \in \{0\} \cup \mathscr{P}^<(x^*).$$

The latter inequalities imply, for every $x \in S$,

$$f^0(x) + \sum_{k \in \mathscr{P}^<(x^*)} \lambda_k f^k(x) \geq f^0(x^*) + \sum_{k \in \mathscr{P}^<(x^*)} \lambda_k f^k(x^*)$$
$$+ (x - x^*)^T \left(y^0 + \sum_{k \in \mathscr{P}^<(x^*)} \lambda_k y^k \right). \tag{3.17}$$

Since for $x \in S$, $f^k(x) \leq 0$, $k \in \mathscr{P}$ and $f^k(x^*) = 0$, $k \in \mathscr{P}^<(x^*) \subset \mathscr{P}(x^*)$ it follows from (3.16) and (3.17) that

$$f^0(x) \geq f^0(x^*) + (x - x^*)^T y, \qquad \forall x \in S. \tag{3.18}$$

By the convexity of S, $x - x^* \in F(S, x^*)$ and therefore, by Lemma 3.5(a), $x - x^* \in D_{\overline{\mathscr{P}}^=}(x^*)$. Since $y \in (D_{\overline{\mathscr{P}}^=}(x^*))^*$, $(x - x^*)^T y \geq 0$, and (3.18) implies $f^0(x) \geq f^0(x^*)$, $\forall x \in S$.

Only if. If x^* is a (unconstrained) minimizer of f^0, then $0 \in \partial f^0(x^*)$ and (3.15) holds with $y = 0$, $\lambda_k = 0$. If x^* is not a minimizer, then by Lemma 2.6

$$(D_0^<(x^*))^* = \left\{ \mu \partial f^0(x^*) : \mu \leq 0 \right\}. \tag{3.19}$$

From Lemma 3.5(c) it follows that x^* is not a minimizer of f^k for each $k \in \mathscr{P}^<(x^*)$. So again Lemma 2.6 applies and

$$(D_k^<(x^*))^* = \left\{ \mu \partial f^k(x^*) : \mu \leq 0 \right\}. \tag{3.20}$$

Specifying (3.19) and (3.20), and using Theorem 3.6, invokes the existence of

$$\mu_0 < 0, \quad \mu_k \le 0, \quad k \in \mathcal{P}^<(x^*)$$

$$y^k \in \partial f^k(x^*), \quad k \in \{0\} \cup \mathcal{P}^<(x^*)$$

$$u \in (D_{\mathcal{P}^-}^=(x^*))^*$$

such that

$$-u = \mu_0 y^0 + \sum_{k \in \mathcal{P}^<(x^*)} \mu_k y^k$$

which implies (3.15) by choosing

$$y = -\frac{u}{\mu_0}, \quad \lambda_k = \frac{\mu_k}{\mu_0}, \quad k \in \mathcal{P}^<(x^*).$$

∎

The characterizations of an optimum in Corollary 3.7 are greatly simplified if

$$\mathcal{P}^= = \varnothing.$$

This condition is equivalent to the so-called *Slater condition*:

$$\exists \hat{x} \quad \text{such that} \quad f^k(\hat{x}) < 0, \quad \forall k \in \mathcal{P}. \tag{3.21}$$

Indeed, when $\mathcal{P}^= = \varnothing$ then $D_{\mathcal{P}^-}^=(x^*) = R^n$, $(D_{\mathcal{P}^-}^=(x^*))^* = \{0\}$, and (3.15) becomes

$$\begin{cases} \text{there exist non-negative scalars } \lambda_k \ge 0, k \in \mathcal{P}(x^*) \\ \text{such that} \\ 0 \in \partial f^0(x^*) + \sum_{k \in \mathcal{P}(x^*)} \lambda_k \partial f^k(x^*). \end{cases} \tag{3.22}$$

This formulation is equivalent to the original *Kuhn-Tucker saddle point condition* (see Ex. 3.13).

The result (3.22) appears in, for example, ROCKAFELLAR [70a] and PSHENICHNYI [71]. In the differentiable case one can use Corollaries 1.5(a) and (2.4) to further simplify Theorem 3.6.

3.8 Corollary. Consider program (P) with differentiable convex functions. A feasible point x^* is optimal if and only if the system

$$\nabla f^0(x^*)^T d < 0$$

$$\nabla f^k(x^*)^T d < 0, \quad k \in \mathcal{P}^<(x^*)$$

$$d \in D_k^=(x^*), \quad k \in \mathcal{P}^=$$

is inconsistent, or dually, if and only if

$$\begin{cases} \text{there exists a vector } y \in (D_{\mathscr{P}^-}^{=}(x^*))^* \\ \text{and non-negative scalars } \lambda_k \geqslant 0, \qquad k \in \mathscr{P}^<(x^*) \\ \text{such that} \\ \nabla f^0(x^*) + \sum_{k \in \mathscr{P}^<(x^*)} \lambda_k \nabla f^k(x^*) = y. \end{cases} \qquad (3.23)$$

■

Theorem 3.6 and Corollaries 3.7 and 3.8 use the set $\mathscr{P}^=$. There remains the problem of determining which of the 2^{p^*} subsets of $\mathscr{P}(x^*)$ is the sought set $\mathscr{P}^=$. This can be done effectively, using at most p^* iterations, by means of the following algorithm.

3.9 Algorithm for constructing $\mathscr{P}^=$ (ABRAMS and KERZNER [78]). This is an iterative method. Starting with $\Omega = \varnothing$, the set Ω is increased at every iteration (except if $\mathscr{P}^= = \varnothing$), and the algorithm terminates when $\Omega = \mathscr{P}^=$ is reached. Termination of the algorithm is based on the properties of $\mathscr{P}^=$ given in Lemma 3.5.

Let x^* be a fixed, but arbitrary, point in S. For a subset Ω of $\mathscr{P}(x^*)$, denote its complement in $\mathscr{P}(x^*)$ by $\text{comp}\,\Omega$. Then solve the system

$(A.\Omega)$

$$y + \sum_{k \in \text{comp}\,\Omega} y^k = 0$$

$$y \in (D_{\Omega}^{=}(x^*))^*, \qquad y^k \in (D_k^<(x^*))^*, \quad k \in \text{comp}\,\Omega, \text{ not all zero.} \quad (3.24)$$

Let

$$\bar{\Omega} \underset{\Delta}{=} \{k \in \text{comp}\,\Omega : y^k \neq 0\}.$$

If $\bar{\Omega} = \varnothing$ [i.e., if $(A.\Omega)$ is inconsistent], then $\Omega = \mathscr{P}^=$. [This happens in the first iteration, that is, when $\Omega = \varnothing$, if and only if Slater's condition (3.21) holds.] Otherwise, the next iteration begins with the new Ω,

$$\Omega_{\text{NEW}} = \Omega \cup \bar{\Omega}. \qquad (3.25)$$

Proof. If the system $(A.\Omega)$ is inconsistent, then by the separation Theorem 2.3 and Lemma 3.5(b)

$$D_{\text{comp}\,\Omega}^<(x^*) \cap \text{conv}\, D_{\Omega}^{=}(x^*) \neq \varnothing,$$

which, by Lemma 3.5(d), proves that $\Omega = \mathscr{P}^=$.

Suppose that (A.Ω) is consistent. We show that

$$\overline{\Omega} \subset \mathscr{P}^=,$$

which guarantees that the set Ω_{NEW}, defined by (3.25), remains in $\mathscr{P}^=$. Indeed, if there exists a $k_0 \in \overline{\Omega}$, $k_0 \notin \mathscr{P}^=$ then it violates (3.11) and there exists a vector d^0 satisfying

$$d^0 \in F(S, x^*) \cap D_{k_0}^{\leqq}(x^*). \tag{3.26}$$

Since $\Omega \subset \mathscr{P}^=$ it follows from (3.26) and Lemma 3.5(a) that

$$d^0 \in D_{k_0}^{\leqq}(x^*) \cap D_{\text{comp}\,\Omega \backslash k_0}^{\leqq}(x^*) \cap D_{\overline{\Omega}}^{=}(x^*).$$

Let $y, \{ y^k : k \in \text{comp}\,\Omega \}$ be a solution of (A.Ω). The inner product of d^0 with (3.24) is

$$y^T d^0 + \sum_{k \in \text{comp}\,\Omega \backslash k_0} (y^k)^T d^0 + (y^{k_0})^T d^0 = 0. \tag{3.27}$$

Then

$$y^T d^0 \geqslant 0, \quad \text{since} \quad d^0 \in D_{\overline{\Omega}}^{=}(x^*), \quad y \in (D_{\overline{\Omega}}^{=}(x^*))^*;$$

similarly

$$(y^k)^T d^0 \geqslant 0, \quad k \in \text{comp}\,\Omega \backslash k_0$$

and

$$(y^{k_0})^T d^0 > 0, \quad \text{by Lemma 2.2(c),}$$

since

$$0 \neq y^{k_0} \in (D_{k_0}^{\leqq}(x^*))^* \quad \text{and} \quad d^0 \in D_{k_0}^{\leqq}(x^*), \quad \text{which is open}.$$

These inequalities contradict (3.27).   ∎

In most practical situations, in particular when the problem functions are differentiable and faithfully convex, the optimality conditions of this section are statements about the solvability of finite systems of *linear* inequalities. Likewise, Algorithm 3.9 involves solving only linear programs (see Exs. 3.5–3.6).

The various characterizations for the differentiable case, in primal and dual form, are summarized in Table 3.1. The results are given for a problem in which the linear and nonlinear constraints are grouped separately:

(PL)

$$\min f^0(x)$$

TABLE 3.1 A Summary of Optimality Conditions in Convex Programming

Let x^* be a feasible solution of

$$(PL) \quad \min\{f^0(x): f^k(x) < 0, \quad k \in \mathcal{R}, \quad x^T a^i - \beta_i < 0, \quad i \in \mathcal{L}\}$$

where $\{f^k : k \in \{0\} \cup \mathcal{R}\}$ are convex nonlinear differentiable functions: $R^n \to R$. Let $\mathcal{R}(x^*) = \{k \in \mathcal{R} : f^k(x^*) = 0\}$, $\mathcal{L}(x^*) = \{i \in \mathcal{L} : (x^*)^T a^i = \beta_i\}$; $\mathcal{R}^= = \{k \in \mathcal{R} : f^k(x) = 0$ for all feasible $x\}$, $\mathcal{R}^<(x^*) = \mathcal{R}(x^*)\backslash\mathcal{R}^=$ and $C^* = \{\sum_{i\in\mathcal{L}(x^*)}\mu_i a^i : \mu_i < 0, \ i \in \mathcal{L}(x^*)\}$.

Name and type	Primal	Dual	Comment
Kuhn–Tucker sufficient condition	There is no $d \in R^n$ such that $\nabla f^0(x^*)^T d < 0$ $\nabla f^k(x^*)^T d < 0, \quad k \in \mathcal{R}(x^*)$ $d^T a^i < 0, \quad i \in \mathcal{L}(x^*)$	There exist $\lambda_k, k \in \mathcal{R}(x^*)$ and $\mu_i, i \in \mathcal{L}(x^*)$ such that $\nabla f^0(x^*) + \sum_{k\in\mathcal{R}(x^*)}\lambda_k \nabla f^k(x^*) + \sum_{i\in\mathcal{L}(x^*)}\mu_i a^i = 0$ $\lambda_k > 0, \quad k \in \mathcal{R}(x^*), \qquad \mu_i > 0, \quad i \in \mathcal{L}(x^*)$	A special assumption called *constraint qualification* is needed for this condition to be necessary (see Exs. 3.9–3.12)
"Multi-Ω" necessary and sufficient condition (a differentiable version of Theorem 3.2)	For every subset Ω of $\mathcal{R}(x^*)$ there is no $d \in R^n$ such that $\nabla f^0(x^*)^T d < 0$ $\nabla f^k(x^*)^T d < 0, \quad k \in \mathcal{R}(x^*)\backslash\Omega$ $d \in D_k^=(x^*), \quad k \in \Omega$ $d^T a^i < 0, \quad i \in \mathcal{L}(x^*)$.	For every subset Ω of $\mathcal{P}(x^*)$ there exist $\lambda_k, k \in \{0\} \cup (\mathcal{R}(x^*)\backslash\Omega)$ such that $\sum_{k\in\{0\}\cup(\mathcal{R}(x^*)\backslash\Omega)}\lambda_k \nabla f^k(x^*) \in \text{cl}(D_{\mathcal{R}^=}^= - (x^*)^* + C^*)$ $\lambda_k > 0, \quad k \in \{0\} \cup (\mathcal{R}(x^*)\backslash\Omega),$ not all zero.	If $f^k, k \in \mathcal{R}^=$ are faithfully convex, the "closure" can be omitted
"Minimal subset" necessary and sufficient condition (Corollary 3.8)	There is no $d \in R^n$ such that $\nabla f^0(x^*)^T d < 0$ $\nabla f^k(x^*)^T d < 0, \quad k \in \mathcal{R}^<(x^*)$ $d \in D_{\mathcal{R}^=}^= - (x^*)$ $d^T a^i < 0, \quad i \in \mathcal{L}(x^*)$.	There exist $\lambda_k, k \in \{0\} \cup \mathcal{R}^<(x^*)$ such that $\nabla f^0(x^*) + \sum_{k\in\mathcal{R}^<(x^*)}\lambda_k \nabla f^k(x^*) \in \text{cl}(D_{\mathcal{R}^=}^= - (x^*)^* + C^*)$ $\lambda_k > 0, \quad k \in \mathcal{R}^<(x^*)$	
Fritz John necessary condition	There is no $d \in R^n$ such that $\nabla f^0(x^*)^T d < 0$ $\nabla f^k(x^*)^T d < 0, \quad k \in \mathcal{R}(x^*)$ $d^T a^i < 0, \quad i \in \mathcal{L}(x^*)$.	There exist $\lambda_k, k \in \{0\} \cup \mathcal{R}(x^*)$, and $\mu_i, i \in \mathcal{L}(x^*)$ such that $\sum_{k\in\{0\}\cup(\mathcal{R}(x^*)}\lambda_k \nabla f^k(x^*) + \sum_{i\in\mathcal{L}(x^*)}\mu_i a^i = 0$ $\lambda_k > 0, \ k \in \{0\} \cup \mathcal{R}(x^*)$ not all zero, $\mu_i > 0, i \in \mathcal{L}(x^*)$.(see Exs. 3.14, 3.15, and 3.17–3.19)	A special assumption called *reduction condition* is needed for this condition to be sufficient

s.t.

$$\text{nonlinear:} \quad f^k(x) \le 0, \quad k \in \mathfrak{N}$$

$$\text{linear:} \quad x^T a^i - \beta_i \le 0, \quad i \in \mathfrak{L}.$$

The binding constraints are denoted by

$$\mathfrak{N}(x^*) = \{ k \in \mathfrak{N} : f^k(x^*) = 0 \}, \quad \mathfrak{L}(x^*) = \{ i \in \mathfrak{L} : (x^*)^T a^i = \beta_i \}.$$

In this case it is easy to see that the proper notion of the minimal index set of binding constraints is

$$\mathfrak{N}^= = \{ k \in \mathfrak{N} : x \in S \Rightarrow f^k(x) = 0 \}$$

where, as before, S is the feasible set of (PL). Note that the difference between $\mathfrak{N}^=$ and $\mathfrak{P}^=$ is the fact that $\mathfrak{N}^=$ contains only the *nonlinear* binding constraints, while $\mathfrak{P}^=$ is the minimal subset of *all* binding constraints: $\mathfrak{N}(x^*) \cup \mathfrak{L}(x^*)$.

Exercises and Examples

3.1   A point with no directions of descent that is not a local minimum.   Let $f: R^2 \to R$ be given by

$$f(x_1, x_2) = (x_2 - x_1^2)(x_2 - 2x_1^2)$$

and consider the point $x^* = (0,0)^T$. Then

(a)   At x^* every nonzero vector is a direction of ascent, that is,

$$D_f^>(x^*) = R^2 \backslash \{0\}.$$

Indeed, let $d = (d_1, d_2)^T \ne 0$. Then

$$f(x^* + \alpha d) = \alpha^2 (d_2 - \alpha d_1^2)(d_2 - 2\alpha d_1^2)$$
$$> 0 = f(x^*), \quad \forall \quad 0 < \alpha < \bar{\alpha}$$

where

$$\bar{\alpha} = \begin{cases} +\infty & \text{if} \quad d_1 = 0 \quad \text{or} \quad d_2 \le 0 \\ \dfrac{d_2}{2d_1^2} & \text{if} \quad d_1 \ne 0 \quad \text{and} \quad d_2 > 0. \end{cases}$$

(b)   The point x^* is not a local minimum for f. Indeed,

$$f(\epsilon, \tfrac{3}{2}\epsilon^2) = -\tfrac{1}{4}\epsilon^2 < 0 = f(x^*) \quad \forall \quad \epsilon.$$

The above example shows that a convexity assumption (such as strict quasiconvexity) on f^0 is essential for Lemma 3.1.

THEOREMS OF ALTERNATIVES

A theorem of alternatives lists two systems, exactly one of which is consistent. Two such theorems are given below.

3.2　Let A be any $m \times n$ real matrix. Then exactly one of the following two systems is consistent (GORDAN [1873]):

(a)　$Au = 0,\ u \geqslant 0,\ u \neq 0,$
(b)　$A^T v < 0.$

From Theorem 3.2 it follows (for differentiable functions and $\Omega = \emptyset$) that at an optimal x^* the system

$$\nabla f^k(x^*)^T d < 0, \qquad k \in \{0\} \cup \mathscr{P}(x^*)$$

is inconsistent. The latter is equivalent to the Fritz John condition:

$$\begin{cases} \text{there exist multipliers } \lambda_k \geqslant 0,\ k \in \{0\} \cup \mathscr{P}(x^*) \text{ not all zero} \\ \text{such that} \\ \displaystyle\sum_{k \in \{0\} \cup \mathscr{P}(x^*)} \lambda_k \nabla f^k(x^*) = 0. \end{cases} \qquad \text{(FJ)}$$

This equivalence can be shown directly by using Gordan's theorem (see also Ex. 3.12).

3.3　Let $A = [a_1, a_2, \ldots, a_n]$ be an $m \times n$ matrix. Then exactly one of the following two systems is consistent:

(a)　$Au = 0,\ u \geqslant 0,\ u_1 > 0,$
(b)　$A^T v \leqslant 0,\ a_1^T v < 0.$

This is a special case of Motzkin's theorem of alternatives, MOTZKIN [36]. An application of this theorem is given in Ex. 3.4.

3.4　Sufficiency of the Kuhn-Tucker Condition.　From Motzkin's theorem it follows that the Kuhn-Tucker condition

$$\begin{cases} \text{there exist multipliers } \lambda_k \geqslant 0,\ k \in \mathscr{P}(x^*) \\ \text{such that} \\ \nabla f^0(x^*) + \displaystyle\sum_{k \in \mathscr{P}(x^*)} \lambda_k \nabla f^k(x^*) = 0 \end{cases} \qquad \text{(K-T)}$$

is equivalent to the inconsistency of the system

$$\nabla f^0(x^*)^T d < 0$$

$$\nabla f^k(x^*)^T d \leqslant 0, \qquad k \in \mathcal{P}(x^*).$$

But, since

$$D_k^{\leqslant}(x^*) \subset \left\{ d : \nabla f^k(x^*)^T d \leqslant 0 \right\}$$

by Corollary 1.5, the latter means that

$$D_0^{<}(x^*) \cap D_{\mathcal{P}(x^*)}^{\leqslant}(x^*) = \varnothing.$$

Hence x^* is optimal, by (1.17) and Lemma 3.1. Therefore the Kuhn-Tucker condition is sufficient for optimality, but in general not necessary, since it is possible that

$$D_k^{\leqslant}(x^*) \neq \left\{ d : \nabla f^k(x^*)^T d \leqslant 0 \right\}.$$

OPTIMALITY CONDITIONS

The faithfully convex case.   In this case, because of Lemma 1.6, all optimality conditions that involve directions of constancy are greatly simplified, and, in fact, reduce to checking systems of linear equations and inequalities. Corollary 3.8, for example, reduces, in the faithfully convex case, to Ex. 3.5.

3.5   Let (P) and x^* be as in Corollary 3.8 and let the functions $\{ f^k : k \in \mathcal{P}^= \}$ be faithfully convex, that is, of the form

$$f^k(x) = h^k\left(A_k x + b^k\right) + \left(a^k\right)^T x + \alpha_k \tag{3.28}$$

where $h^k : R^{m_k} \to R$ is strictly convex, $A_k \in R^{m_k \times n}$, $b^k \in R^{m_k}$, $a^k \in R^n$, $\alpha_k \in R$ for $k \in \mathcal{P}^=$. Then x^* is optimal if and only if the system

$$\nabla f^0(x^*)^T d < 0$$

$$\nabla f^k(x^*)^T d < 0, \qquad k \in \mathcal{P}^<(x^*)$$

$$\left.\begin{matrix} A_k d = 0 \\ \left(a^k\right)^T d = 0 \end{matrix}\right\} k \in \mathcal{P}^=$$

is inconsistent, or dually, if and only if the following system (with unknowns $\{\lambda_k : k \in \mathcal{P}^<(x^*)\}$ and $\{u^k \in R^{m_k+1} : k \in \mathcal{P}^=\}$) is consistent:

$$\nabla f^0(x^*) + \sum_{k \in \mathcal{P}^<(x^*)} \lambda_k \nabla f^k(x^*) = \sum_{k \in \mathcal{P}^=} \left[A_k^T, a^k \right] u^k$$

$$\lambda_k \geqslant 0, \qquad k \in \mathcal{P}^<(x^*).$$

Proof. Follows from Corollary 3.8 once we show that for the faithfully convex functions (3.28),

$$(D_{\mathcal{P}^-}^=(x^*))^* = \sum_{k \in \mathcal{P}^-} [A_k^T, a^k] u^k, \qquad u^k \in R^{m_k+1}. \qquad (3.29)$$

This follows from

$$(D_{\mathcal{P}^-}^=(x^*))^* = \text{cl} \sum_{k \in \mathcal{P}^-} (D_k^=(x^*))^*, \qquad \text{by Lemma 2.2}(g)$$

$$= \text{cl} \sum_{k \in \mathcal{P}^-} \left\{ N \left[\begin{bmatrix} A_k \\ (a^k)^T \end{bmatrix} \right] \right\}^*, \qquad \text{by (1.13)}$$

$$= \text{cl} \sum_{k \in \mathcal{P}^-} R([A_k^T, a^k]), \qquad \begin{array}{l} \text{because for any matrix } B, \\ (N(B))^* = (N(B))^\perp, \\ \text{by Lemma 2.2}(e) \\ = R(B^T) \end{array}$$

$$= \sum_{k \in \mathcal{P}^-} R([A_k^T, a^k]), \qquad \begin{array}{l} \text{since the sum of subspaces} \\ \text{of } R^n \text{ is a subspace of } R^n, \\ \text{hence closed,} \end{array}$$

which proves (3.29). ∎

Algorithm 3.9. The differentiable, faithfully convex case. In this case, Algorithm 3.9 for constructing $\mathcal{P}^=$ is greatly simplified: At each iteration one solves a system of linear (homogeneous) equations, with certain variables restricted to be non-negative. This is described in Ex. 3.6 and illustrated in Ex. 3.7.

3.6 Algorithm for constructing $\mathcal{P}^=$ in the differentiable, faithfully convex case. Let

$$S = \{ x : f^k(x) \leq 0, \qquad k \in \mathcal{P} \}$$

where the functions f^k, $k \in \mathcal{P}$, are differentiable and faithfully convex:

$$f^k(x) = h^k(A_k x + b^k) + (a^k)^T x + \alpha_k. \qquad (3.28)$$

Initialization. Choose any $x^* \in S$ and set $\Omega = \emptyset$.

Step 1 Solve the system (with unknowns $\{\lambda_k : k \in \mathcal{P}(x^*) \setminus \Omega\}$ and $\{u^k \in R^{m_k+1} : k \in \Omega\}$)

(A.Ω)

$$\sum_{k \in \mathscr{P}(x^*)\backslash\Omega} \lambda_k \nabla f^k(x^*) - \sum_{k \in \Omega} \left[A_k^T, a^k \right] u^k = 0$$

$$\lambda_k \geqslant 0, \qquad \sum_{k \in \mathscr{P}(x^*)\backslash\Omega} \lambda_k = 1.$$

Step 2 If (A.Ω) is inconsistent, terminate with $\mathscr{P}^= = \Omega$.

Step 3 If (A.Ω) is consistent, set

$$\overline{\Omega} = \{ k \in \mathscr{P}(x^*)\backslash\Omega : \lambda_k > 0 \},$$

$$\Omega_{\text{NEW}} = \Omega \cup \overline{\Omega}$$

and repeat step 1 with $\Omega = \Omega_{\text{NEW}}$.

3.7 Example (Algorithm 3.9). Let $S \subset R^5$ be defined by the constraints

$$
\begin{aligned}
f^1(x) &= e^{x_1} + x_2^2 && - 1 \leqslant 0 \\
f^2(x) &= x_1^2 + x_2^2 + e^{-x_3} && - 1 \leqslant 0 \\
f^3(x) &= x_1 && + x_4^2 + x_5^2 - 1 \leqslant 0 \\
f^4(x) &= e^{-x_2} && - 1 \leqslant 0 \\
f^5(x) &= (x_1 - 1)^2 + x_2^2 && - 1 \leqslant 0 \\
f^6(x) &= x_1 && + e^{-x_4} \quad - 1 \leqslant 0 \\
f^7(x) &= x_2 && + e^{-x_5} - 1 \leqslant 0.
\end{aligned}
$$

$$(3.30)$$

Initialization. We choose $x^* = (0, 0, 1, \frac{1}{2}\sqrt{2}, \frac{1}{2}\sqrt{2})^T$. Then the binding constraints are $\mathscr{P}(x^*) = \{1, 3, 4, 5\}$, with the corresponding gradients

$$\nabla f^1(x^*) = (1, 0, 0, 0, 0)^T$$

$$\nabla f^3(x^*) = (1, 0, 0, \sqrt{2}, \sqrt{2})^T$$

$$\nabla f^4(x^*) = (0, -1, 0, 0, 0)^T$$

$$\nabla f^5(x^*) = (-2, 0, 0, 0, 0)^T.$$

Set $\Omega = \varnothing$.

Iteration 1

Step 1 Solve

(A.Ø)

$$\lambda_1 \begin{bmatrix} 1 \\ 0 \\ 0 \\ 0 \\ 0 \end{bmatrix} + \lambda_3 \begin{bmatrix} 1 \\ 0 \\ 0 \\ \sqrt{2} \\ \sqrt{2} \end{bmatrix} + \lambda_4 \begin{bmatrix} 0 \\ -1 \\ 0 \\ 0 \\ 0 \end{bmatrix} + \lambda_5 \begin{bmatrix} -2 \\ 0 \\ 0 \\ 0 \\ 0 \end{bmatrix} = \begin{bmatrix} 0 \\ 0 \\ 0 \\ 0 \\ 0 \end{bmatrix}$$

$$\lambda_1 + \lambda_3 + \lambda_4 + \lambda_5 = 1$$
$$\lambda_1, \quad\quad \lambda_3, \quad\quad \lambda_4, \quad\quad \lambda_5 \geqslant 0.$$

The solution is

$$\lambda_1 = \tfrac{2}{3}, \quad \lambda_3 = 0 = \lambda_4, \quad \lambda_5 = \tfrac{1}{3}.$$

Step 3

$$\overline{\Omega} = \{1,5\}$$
$$\Omega_{\text{NEW}} = \emptyset \cup \overline{\Omega} = \{1,5\}.$$

Iteration 2

Step 1 Solve

(A.$\{1,5\}$)

$$\lambda_3 \begin{bmatrix} 1 \\ 0 \\ 0 \\ \sqrt{2} \\ \sqrt{2} \end{bmatrix} + \lambda_4 \begin{bmatrix} 0 \\ -1 \\ 0 \\ 0 \\ 0 \end{bmatrix} - \begin{bmatrix} 1 & 0 \\ 0 & 1 \\ 0 & 0 \\ 0 & 0 \\ 0 & 0 \end{bmatrix} u^1 - \begin{bmatrix} 1 & 0 \\ 0 & 1 \\ 0 & 0 \\ 0 & 0 \\ 0 & 0 \end{bmatrix} u^5 = \begin{bmatrix} 0 \\ 0 \\ 0 \\ 0 \\ 0 \end{bmatrix}$$

$$\lambda_3 + \lambda_4 = 1$$
$$\lambda_3, \lambda_4 \geqslant 0.$$

A solution is

$$\lambda_3 = 0, \quad \lambda_4 = 1, \quad u^1 = \begin{bmatrix} 0 \\ -\tfrac{1}{2} \end{bmatrix}, \quad u^5 = \begin{bmatrix} 0 \\ -\tfrac{1}{2} \end{bmatrix}.$$

Step 3

$$\overline{\Omega} = \{4\}$$
$$\Omega_{\text{NEW}} = \{1, 4, 5\}.$$

Iteration 3

Step 1　Solve

(A.{1, 4, 5})

$$
\lambda_3 \begin{bmatrix} 1 \\ 0 \\ 0 \\ \sqrt{2} \\ \sqrt{2} \end{bmatrix} - \begin{bmatrix} 1 & 0 \\ 0 & 1 \\ 0 & 0 \\ 0 & 0 \\ 0 & 0 \end{bmatrix} u^1 - \begin{bmatrix} 0 \\ 1 \\ 0 \\ 0 \\ 0 \end{bmatrix} u^4 - \begin{bmatrix} 1 & 0 \\ 0 & 1 \\ 0 & 0 \\ 0 & 0 \\ 0 & 0 \end{bmatrix} u^5 = \begin{bmatrix} 0 \\ 0 \\ 0 \\ 0 \\ 0 \end{bmatrix}
$$

$$\lambda_3 = 1$$
$$\lambda_3 \geqslant 0$$

This system is inconsistent.

Step 2

$$\mathscr{P}^= = \{1, 4, 5\}, \text{ stop.}$$

∎

3.8　Example (Corollary 3.8).　Consider the problem:

$$(P) \quad \min f^0(x) = x_1 - x_2 + (x_3 - 1)^2 + (x_4 - 2)^2 + (x_5 - 2)^2$$

subject to the constraints (3.30) of Ex. 3.7. We check the optimality of

$$x^* = \left(0, 0, 1, \tfrac{1}{2}\sqrt{2}, \tfrac{1}{2}\sqrt{2}\right)^T,$$

which, by Corollary 3.8, is equivalent to the consistency of the system (3.23). But $\mathscr{P}^= = \{1, 4, 5\}$ was computed in Ex. 3.7. So the system to be checked is

$$
\begin{bmatrix} 1 \\ -1 \\ 0 \\ \sqrt{2} - 4 \\ \sqrt{2} - 4 \end{bmatrix} + \lambda_3 \begin{bmatrix} 1 \\ 0 \\ 0 \\ \sqrt{2} \\ \sqrt{2} \end{bmatrix} = \begin{bmatrix} 1 & 0 \\ 0 & 1 \\ 0 & 0 \\ 0 & 0 \\ 0 & 0 \end{bmatrix} u^1 + \begin{bmatrix} 0 \\ 1 \\ 0 \\ 0 \\ 0 \end{bmatrix} u^4 + \begin{bmatrix} 1 & 0 \\ 0 & 1 \\ 0 & 0 \\ 0 & 0 \\ 0 & 0 \end{bmatrix} u^5
$$

$$\lambda_3 \geqslant 0;$$

since it is consistent (with $\lambda_3 = 2\sqrt{2} - 1$, $u^1 = (2\sqrt{2}, 0)^T$, $u^4 = -1$, $u^5 = 0$), the point x^* is optimal.

This optimal x^* does not satisfy the Kuhn-Tucker condition:

$$
\begin{bmatrix} 1 \\ -1 \\ 0 \\ \sqrt{2} - 4 \\ \sqrt{2} - 4 \end{bmatrix} + \lambda_1 \begin{bmatrix} 1 \\ 0 \\ 0 \\ 0 \\ 0 \end{bmatrix} + \lambda_3 \begin{bmatrix} 1 \\ 0 \\ 0 \\ \sqrt{2} \\ \sqrt{2} \end{bmatrix} + \lambda_4 \begin{bmatrix} 0 \\ -1 \\ 0 \\ 0 \\ 0 \end{bmatrix} + \lambda_5 \begin{bmatrix} -2 \\ 0 \\ 0 \\ 0 \\ 0 \end{bmatrix} = \begin{bmatrix} 0 \\ 0 \\ 0 \\ 0 \\ 0 \end{bmatrix}
$$

$$
\lambda_1, \lambda_3, \lambda_4, \lambda_5 \geqslant 0.
$$

CONSTRAINT QUALIFICATION

In a convex programming problem, a *constraint qualification* is a condition under which the Kuhn-Tucker condition is necessary for optimality (for any objective function). Examples of constraint qualification are given in Exs. 3.9 and 3.10.

3.9 A local constraint qualification. Let x^* be a feasible solution of a convex program (P) with differentiable constraint functions. Assume that the gradients of the binding constraints at x^* are non-negatively linearly independent, that is, there are no non-negative $\{\lambda_k : k \in \mathcal{P}(x^*)\}$ not all zero, such that

$$
\sum_{k \in \mathcal{P}(x^*)} \lambda_k \nabla f^k(x^*) = 0.
$$

In this case we say that x^* is *regular*.

A regular x^* is optimal if and only if it satisfies the Kuhn-Tucker condition.

Proof. Let x^* be regular and optimal. Then it satisfies the Fritz John condition (FJ). Since $\lambda_0 = 0$ contradicts the assumed regularity of x^*, λ_0 is positive. Division of the equation by λ_0 shows that the Kuhn-Tucker condition is satisfied with multipliers

$$
\frac{\lambda_k}{\lambda_0}, \qquad k \in \mathcal{P}(x^*).
$$

∎

THE SLATER CONDITION

The most popular constraint qualification is the *Slater condition*: There exists a point $\hat{x} \in R^n$ such that

$$f^k(\hat{x}) < 0, \qquad \forall k \in \mathcal{P}. \tag{3.31}$$

This condition is a global constraint qualification, in the sense that it is a characteristic of the feasible set as a whole (see also Ex. 3.12 below).

The Slater condition is occasionally referred to in the literature as an *interiority condition*. This term is misleading, since the feasible set may have a nonempty interior, yet violate the Slater condition, for example,

$$S = \{x \in R : 0 \cdot x \leqslant 0\} = R.$$

3.10 Let the convex program (P) satisfy the Slater condition, and let x^* be an optimal solution of (P). If $\{f^k : k \in \{0\} \cup \mathcal{P}(x^*)\}$ are differentiable, then the Kuhn-Tucker condition is satisfied at x^*.

Proof. Since Slater's condition is equivalent to $\mathcal{P}^= = \emptyset$, the result is an immediate consequence of Corollary 3.8.   ∎

3.11 Checking the Slater condition. Algorithm 3.9 checks the Slater condition in one iteration.

3.12 Let

$$S = \left\{ x : f^k(x) \leqslant 0, \qquad k \in \mathcal{P} \right\}$$

where the functions $\{f^k : k \in \mathcal{P}\}$ are convex and differentiable, and let $x^* \in S$. Then S satisfies the Slater condition if and only if x^* is regular.

Proof. The regularity of x^*, that is, the inconsistency of the system

$$\begin{cases} \sum_{k \in \mathcal{P}(x^*)} \lambda_k \nabla f^k(x^*) = 0 \\ \lambda_k \geqslant 0, \qquad k \in \mathcal{P}(x^*), \qquad \text{not all} \quad \lambda_k = 0 \end{cases}$$

is equivalent, by Ex. 3.2, to the existence of $\hat{d} \in R^n$, satisfying

$$\nabla f^k(x^*)^T \hat{d} < 0, \qquad k \in \mathcal{P}(x^*)$$

which, by Corollary 1.5(a) and the fact that $f^k(x^*) < 0$ for $k \in \mathcal{P} \setminus \mathcal{P}(x^*)$, is equivalent to the existence of $\hat{x} = x^* + \alpha \hat{d}$, $\alpha > 0$ sufficiently small, satisfying (3.31).   ∎

A SADDLE POINT CHARACTERIZATION

Condition (3.22), which is necessary and sufficient for optimality under Slater's constraint qualification, is equivalent (see, for example, BEN-TAL and BEN-ISRAEL [79]) to the classical saddle point characterization of KUHN and TUCKER [51].

3.13 Kuhn-Tucker saddle point theorem. If the convex program (P) satisfies Slater's condition, then a feasible point $x^* \in S$ is optimal if and only if there exists $\lambda^* = (\lambda_k^*) \geqslant 0$ such that $\lambda_k^* f^k(x^*) = 0$, $k \in \mathcal{P}$, and (x^*, λ^*) is a saddle point of the Lagrangian

$$L(x, \lambda) \triangleq f^0(x) + \sum_{k \in \mathcal{P}} \lambda_k f^k(x)$$

that is,

$$L(x^*, \lambda) \leqslant L(x^*, \lambda^*) \leqslant L(x, \lambda^*) \qquad \forall x \in R^n, \qquad \forall \lambda \geqslant 0.$$

■

STRICTLY CONVEX RESTRICTIONS

In this case, by using Ex. 1.9, all the optimality conditions given above are significantly simplified. For example, the primal condition in Theorem 3.2 reduces to the following.

3.14 Let (P) and x^* be as in Theorem 3.2. Assume also that the functions $\{f^k : k \in \{0\} \cup \mathcal{P}\}$ are differentiable and $\{f^k : k \in \mathcal{P}(x^*)\}$ have strictly convex restrictions $f^{[k]}$. Then it follows that x^* is optimal if and only if for every subset Ω of $\mathcal{P}(x^*)$ the following system (of linear inequalities and equations) is inconsistent:

$$\nabla f^0(x^*)^T d < 0$$
$$\nabla f^k(x^*)^T d < 0, \qquad k \in \mathcal{P}(x^*) \backslash \Omega \qquad (3.32)$$
$$d_{[k]} = 0, \qquad k \in \Omega.$$

■

REDUCTION CONDITIONS

For problem (P) with differentiable functions, a *reduction condition* at a feasible point x^* guarantees that the Fritz John condition is not only necessary, but also *sufficient* for optimality of x^*. (We encounter such a condition earlier in Corollary 3.3.)

3.15 Let (P), x^*, and $\{f^k : k \in \{0\} \cup \mathscr{P}\}$ be as in Ex. 3.14. Then a reduction condition at x^* is

$$\left\{ j : \frac{\partial}{\partial x_j} f^0(x^*) \neq 0 \right\} \subset [k] \qquad \forall k \in \mathscr{P}(x^*). \tag{R1}$$

Proof. Choose an $\Omega \subset \mathscr{P}(x^*)$, $\Omega \neq \varnothing$, and $k_0 \in \Omega$. Then in system (3.32),

$$d_j = 0 \qquad \forall j \in [k_0].$$

Hence, by (R1), $\nabla f^0(x^*)^T d = 0$. This shows that (3.32) is inconsistent whenever $\Omega \neq \varnothing$. In other words, x^* is optimal if and only if the single system (3.32) with $\Omega = \varnothing$ is inconsistent. But this is equivalent to (FJ). ∎

3.16 Example for the reduction condition (R1). Consider

$$\min f^0 = x_1^2 + e^{-x_2} + x_3^2$$

s.t.

$$f^1 = e^{x_1} + x_2^2 \quad - 1 \leqslant 0$$
$$f^2 = \quad e^{-x_2} \quad - 1 \leqslant 0$$
$$f^3 = (x_1 - 1)^2 + x_2^2 \quad - 1 \leqslant 0$$
$$f^4 = x_1^2 + x_2^2 + e^{-x_3} \quad - 1 \leqslant 0$$

Since (R1) holds at $x^* = 0$, this point is optimal if and only if (FJ) holds, which indeed establishes its optimality. Let us note that the Kuhn-Tucker condition [i.e., (FJ) with $\lambda_0 = 1$] does not hold here at the optimal x^*.

In the case of programs (P) with linear constraints, Corollary 3.8 assumes the form given in Ex. 3.5.

3.17 Problems with linear constraints. Consider

$$\min f^0(x)$$

s.t.

$$f^k(x) \leqslant 0, \qquad k \in \mathscr{P}$$
$$x^T a^i - \beta_i \leqslant 0, \qquad i \in \mathscr{Q}$$
$$x^T a^j - \beta_j = 0, \qquad j \in \mathscr{Q}_0$$

where $\{f^k : k \in \{0\} \cup \mathscr{P}\}$ are differentiable convex (nonlinear) functions, $\{a^i : i \in \mathscr{Q} \cup \mathscr{Q}_0\}$ are vectors in R^n, and $\mathscr{P}$, $\mathscr{Q}$, $\mathscr{Q}_0$ are finite index sets. Then a feasible point x^* is optimal if and only if for every subset Ω of $\mathscr{P}(x^*)$, the system

$$\nabla f^0(x^*)^T d < 0$$

$$\nabla f^k(x^*)^T d < 0, \qquad k \in \mathcal{P}(x^*)\backslash \Omega$$

$$d \in D_{\bar{\Omega}}^{=}(x^*) \tag{3.33}$$

$$d^T a^i \leqslant 0, \qquad i \in \mathcal{Q}^* \underline{\Delta}\{i \in \mathcal{Q} : (x^*)^T a^i = \beta_i\}$$

$$d^T a^j = 0, \qquad j \in \mathcal{Q}_0$$

is inconsistent.

3.18 Reduction condition for programs with linear constraints. Consider program (P) with linear equalities

$$\min f^0(x)$$

s.t.

$$f^k(x) \leqslant 0, \qquad k \in \mathcal{P}$$

$$Ax = b$$

where $A \in R^{m \times n}$ and $b \in R^m$. Assume that the functions $\{ f^k : k \in \{0\} \cup \mathcal{P}\}$ are differentiable and $\{ f^k : k \in \mathcal{P}(x^*)\}$ have strictly convex restrictions $f^{[k]}$. Then the following is a reduction condition at a feasible x^*.

$$\begin{cases} \text{The columns of } A \text{ that correspond to the indices} \\ \{1, 2, \ldots, n\} \backslash [k], \qquad k \in \mathcal{P} \\ \text{are linearly independent.} \end{cases} \tag{R2}$$

Proof. Let $\Omega \neq \varnothing$ and $k_0 \in \Omega$. Then (3.33) contains the relations

$$d_j = 0, \qquad j \in [k_0]$$

$$d^T A = 0.$$

Hence

$$\sum_{j \notin [k_0]} d_j a^j = 0$$

where a^j is the jth column of A. But (R2) now implies $d_j = 0, j \in [k_0]$. So $d = 0$, violating the inequality $\nabla f^0(x^*)^T d < 0$ in (3.33). ∎

3.19 Example for the reduction condition (R2). Consider

$$\min f^0(x_1, x_2, x_3, x_4)$$

s.t.

$$f^1(x_1, x_2, x_3, x_4) \quad \leqslant 0$$
$$f^2(x_1, x_2, x_4) \quad \leqslant 0$$
$$f^3(x_2, x_3, x_4) \quad \leqslant 0$$
$$f^4(x_1, x_4) \quad \leqslant 0$$
$$f^5(x_2, x_3, x_4) \quad \leqslant 0$$
$$x_1 \quad\quad + x_3 + x_4 \quad = \beta_1$$
$$x_2 + x_3 + 2x_4 = \beta_2$$

where $f^1, \ldots, f^5$ are functions with strictly convex restrictions, but are otherwise *arbitrary*, and β_1, β_2 are arbitrary reals. Here the reduction condition (R2) holds at every feasible point. This follows from the fact that

$$\operatorname{card}(\{1,2,3,4\} \setminus [k]) = 4 - \operatorname{card}[k] \leqslant 2, \quad k = 1, 2, \ldots, 5$$

and that any two columns of the matrix

$$A = \begin{bmatrix} 1 & 0 & 1 & 1 \\ 0 & 1 & 1 & 2 \end{bmatrix},$$

corresponding to the linear equalities of the above problem, are linearly independent.

3.20 For problems (P) with faithfully convex constraint functions of the form (3.28), the reduction condition (R1) can be generalized to read:

$$\nabla f^0(x^*) \in R([A_k^T, a^k]) \quad \forall \; k \in \mathscr{P}(x^*). \tag{R3}$$

3.21 Uniqueness of an optimal solution. An optimal solution x^* of problem (P) is unique if and only if

$$D_0^=(x^*) \cap D_{\mathscr{P}(x^*)}^{\leqslant}(x^*) \cap D_{\mathscr{P}^-}^=(x^*) = \{0\}.$$

∎

3.22 Consider

$$\min f^0(x) = \begin{cases} -\sqrt{x}, & \text{if } x \geqslant 0 \\ \infty & \text{otherwise} \end{cases}$$

s.t.

$$f^1(x) = x \leqslant 0.$$

Both functions are continuous (on their domains). Since, at the optimal solution x^*,

$$D_0^{\leqslant}(x^*) = \{x : x > 0\}, \qquad (D_0^{\leqslant}(x^*))^* = \{x : x \geqslant 0\}$$

and

$$D_1^{\leqslant}(x^*) = \{x : x < 0\}, \qquad (D_1^{\leqslant}(x^*))^* = \{x : x \leqslant 0\}$$

the statements of Theorem 3.6 are satisfied. But since $\partial f^0(x^*) = \emptyset$, the statement of Corollary 3.7 is not satisfied. This example shows that the condition

$$x^* \in \bigcap_{k \in \{0\} \cup \mathcal{P}^{<}(x^*)} \operatorname{int} \operatorname{dom} f^k$$

is needed for the conclusion of Corollary 3.7.

4   A PARAMETRIC APPROACH TO OPTIMALITY CONDITIONS

The optimality theory developed in Section 3 is based on the fact that for a convex function f, a feasible direction is either a direction of descent or a direction of constancy, that is,

$$D_f^{\leqslant}(x^*) = D_f^{<}(x^*) \cup D_f^{=}(x^*).$$

For a convex program this requires consideration of all 2^{p^*} [$p^* =$ card $\mathcal{P}(x^*)$] subsets of $\mathcal{P}(x^*)$ (e.g., Theorem 3.2) or the one subset $\mathcal{P}^=$ (Theorem 3.6), which however had to be calculated.

In this section we try a parametric approach, which is more suitable for numerical methods. The approach is based on the idea of expressing

$$d \in D_f^{<}(x)$$

in a simple way. This is the content of the following lemma.

4.1   Lemma.   Let f be a convex function: $R^n \to R$, $x_0 \in \operatorname{dom} f$ be a point at which f is differentiable, and $\phi : R^n \to R$ be a *positive definite function*, that is, a function satisfying

$$\phi(x) \geqslant 0 \qquad \forall x \in R^n$$

$$\phi(x) = 0 \qquad \text{only if} \quad x = 0,$$

and let $d \in R^n$. Then $d \in D_f^{<}(x_0)$ if and only if there exist a scalar $\theta > 0$ and a vector $u \in D_f^{=}(x_0)$ satisfying

$$\nabla f(x_0)^T d + \theta \phi(d - u) \leqslant 0. \tag{4.1}$$

Proof.   If.   Let d satisfy (4.1) for some $\bar{\theta} > 0$ and $\bar{u} \in D_f^{=}(x_0)$. Then

$$\nabla f(x_0)^T d \leqslant -\bar{\theta} \phi(d - \bar{u}) \leqslant 0$$

so that

$$\nabla f(x_0)^T d = 0 \Rightarrow \phi(d - \bar{u}) = 0$$
$$\Rightarrow d = \bar{u}(\in D_f^=(x_0))$$

which by Lemma 1.3(b) proves that

$$d \in D_f^{\leqq}(x_0).$$

Only if. Let $d \in D_f^{\leqq}(x_0)$. Then either $\nabla f(x_0)^T d < 0$, in which case (4.1) is satisfied with

$$u = 0, \qquad \theta = -\frac{\nabla f(x_0)^T d}{\phi(d)}$$

or $\nabla f(x_0)^T d = 0$, in which case $d \in D_f^=(x_0)$, by Lemma 1.3(b), and (4.1) is satisfied with $u = d$ and any $\theta > 0$.   ∎

A natural choice for the positive definite function ϕ in Lemma 4.1 is the ℓ_1-norm:

$$\|x\|_1 = \sum_{j=1}^n |x_j|$$

for which (4.1) reduces to a linear inequality in non-negative variables.

The optimality condition of Lemma 3.1 can now be restated for differentiable and convex objective function and constraints, using Theorem 1.7 and Lemma 4.1.

4.2 Theorem. Let x^* be a feasible solution of program (P). Then x^* is optimal if and only if there exists a positive scalar $\bar{\theta}$ such that for all $\theta \in (0, \bar{\theta}]$ the program (G.θ) has optimal value zero.

(G.θ)

$$\min \nabla f^0(x^*)^T d$$

s.t.

$$\nabla f^k(x^*)^T d + \theta \|d - \delta^k\|_1 \leqslant 0, \qquad k \in \mathcal{P}(x^*)$$
$$\delta^k \in D_k^=(x^*), \qquad k \in \mathcal{P}(x^*). \tag{4.2}$$

∎

Normalizing conditions such as

$$|d_i| \leqslant 1, \qquad i = 1, 2, \ldots, n$$

can be added to the constraints of (G.θ) to bound its optimal value.

For $\theta \geqslant 0$ let $d(\theta)$, $\{\delta^k(\theta) : k \in \mathcal{P}(x^*)\}$ denote an optimal solution of

$(G.\theta)$, and let the optimal value of $(G.\theta)$ be denoted by

$$z(\theta) \underline{\Delta} \nabla f^0(x^*)^T d(\theta).$$

Then, from the way $(G.\theta)$ depends on θ, it is clear that $z(\theta)$ is a monotone nondecreasing function of θ, that is,

$$0 \leqslant \theta_1 < \theta_2 \Rightarrow z(\theta_1) \leqslant z(\theta_2).$$

Theorem 4.2 can therefore be restated as follows:

$$x^* \text{ is optimal if and only if } \liminf_{\theta \to 0^+} z(\theta) = 0.$$

Exercises and Examples

A DUAL SUFFICIENT CONDITION OF THE PARAMETRIC TYPE

The sufficiency condition of Theorem 4.2 can be put into a dual form, as shown in the following theorem.

4.1 A sufficient condition for optimality of a feasible point x^* of (P) is the existence of a positive scalar θ^* such that for every $\theta \in (0, \theta^*]$

$(DG.\theta)$

$$\begin{cases} \text{there exist multipliers } \lambda_k \geqslant 0, \quad k \in \mathcal{P}(x^*) \text{ such that} \\ \nabla f^0(x^*) + \sum_{k \in \mathcal{P}(x^*)} \lambda_k \nabla f^k(x^*) \in \theta \sum_{k \in \mathcal{P}(x^*)} \{(D_k^=(x^*))^* \cap I(\lambda_k)\}. \end{cases}$$

Here $I(\lambda_k) \underline{\Delta} \{x \in R^n : |x_i| \leqslant \lambda_k, \quad i = 1, \ldots, n\}$.

Proof. First we claim that the system

$$\left. \begin{array}{c} \nabla f^0(x^*)^T d < 0 \\ \nabla f^k(x^*)^T d + \theta \sum_{i=1}^n u_i^k + \theta \sum_{i=1}^n v_i^k + w_k = 0 \\ -d + \delta^k + u^k - v^k = 0 \\ \delta^k \in D_k^=(x^*) \\ u^k \geqslant 0, v^k \geqslant 0, w_k \geqslant 0 \end{array} \right\} k \in \mathcal{P}(x^*) \qquad (4.3)$$

has no solution $(d, \delta, u, v, w) \in R^n \times R^{n \times p^*} \times R^{n \times p^*} \times R^{n \times p^*} \times R^{p^*}$. To verify this statement we put (4.3) into the form

$$c^T z < 0, \qquad Az = 0, \qquad z \in S \tag{4.4}$$

where $z = (d, \delta, u, v, w)$, $c = (\nabla f^0(x^*)^T, 0, 0, 0, 0)$,

$$S = R^n \times \left[\underset{k \in \mathcal{P}(x^*)}{\times} D_k^=(x^*) \right] \times R_+^{n \times p^*} \times R_+^{n \times p^*} \times R_+^{p^*},$$

and A is the matrix corresponding to the equations in (4.3). If (4.4) had a solution, then Corollary 2.4 would imply that the system

$$A^T y + c \in S^* \tag{4.5}$$

is inconsistent. Since

$$S^* = \{0\} \times \left\{ \underset{k \in \mathcal{P}(x^*)}{\times} (D_k^=(x^*))^* \right\} \times R_+^{n \times p^*} \times R_+^{n \times p^*} \times R_+^{p^*},$$

the latter implies the inconsistency of

$$\nabla f^0(x^*) + \sum_{k \in \mathcal{P}(x^*)} \lambda_k \nabla f^k(x^*) - \sum_{k \in \mathcal{P}(x^*)} y^k = 0$$

$$\theta \lambda_k \begin{bmatrix} 1 \\ \vdots \\ 1 \end{bmatrix} + y^k \geqslant 0, \qquad \theta \lambda_k \begin{bmatrix} 1 \\ \vdots \\ 1 \end{bmatrix} - y^k \geqslant 0$$

$$\lambda_k \geqslant 0, \qquad y^k \in (D_k^=(x^*))^*, \qquad k \in \mathcal{P}(x^*)$$

in the variables λ_k, y^k. But this is the same as the inconsistency of (DG.θ), a contradiction of the assumption of the theorem.

Next we show that the system

$$\nabla f^0(x^*)^T d < 0$$

$$\nabla f^k(x^*)^T d + \theta \sum_{i=1}^n |d_i - \delta_i^k| \leqslant 0 \tag{4.6}$$

$$\delta^k \in D_k^=(x^*), \qquad k \in \mathcal{P}(x^*)$$

has no solution. Indeed, if $\bar{d}, \bar{\delta}^k, k \in \mathcal{P}(x^*)$ were a solution, then $\bar{d}, \bar{\delta}, \bar{u}, \bar{v}, \bar{w}$, with $\bar{u}, \bar{v},$ and $\bar{w}$ given according to components by

$$\bar{u}_i^k = \tfrac{1}{2} \left(|\bar{d}_i - \bar{\delta}_k^i| + \bar{d}_i - \bar{\delta}_i^k \right)$$

$$\bar{v}_i^k = \tfrac{1}{2} \left(|\bar{d}_i - \bar{\delta}_i^k| - \bar{d}_i + \bar{\delta}_i^k \right), \qquad i = 1, 2, \ldots, n$$

$$\bar{w}_k = -\nabla f^k(x^*)^T \bar{d} - \theta \sum_{i=1}^n \left(\bar{u}_i^k + \bar{v}_i^k \right), \qquad k \in \mathcal{P}(x^*)$$

is a solution of (4.3), a contradiction to its already proven inconsistency. Finally, the fact that (4.6) has no solution is equivalent to the value of the program (G.θ) in Theorem 4.2 being zero. ∎

Condition (DG.θ) contains the Kuhn-Tucker sufficiency criterion, since the zero vector is always contained in

$$\sum_{k \in \mathcal{P}(x^*)} \{(D_k^=(x^*))^* \cap I(\lambda_k)\}.$$

In the example below, the condition (DG.θ) establishes the optimality while the Kuhn-Tucker condition does not.

4.2 Consider

$$\min f^0 = e^{x_1} + e^{-x_2} + x_3$$

s.t.

$$
\begin{aligned}
f^1 &= e^{x_1} && - 1 \leqslant 0 \\
f^2 &= e^{-x_2} && - 1 \leqslant 0 \\
f^3 &= (x_1 - 1)^2 + x_2^2 && - 1 \leqslant 0 \\
f^4 &= x_1^2 + x_2^2 + e^{-x_3} && - 1 \leqslant 0.
\end{aligned}
$$

The point $x^* = (0,0,0)^T$ is tested for optimality. It is easily verified that the Kuhn-Tucker condition [i.e., (DG.θ) for $\theta = 0$] is not satisfied at x^*. On the other hand, (DG.θ) is here

$$-\theta \begin{bmatrix} \lambda_1 + \lambda_3 + \lambda_4 \\ \lambda_2 + \lambda_3 + \lambda_4 \\ \lambda_4 \end{bmatrix} \leqslant \begin{bmatrix} 1 \\ -1 \\ 1 \end{bmatrix} + \lambda_1 \begin{bmatrix} 1 \\ 0 \\ 0 \end{bmatrix} + \lambda_2 \begin{bmatrix} 0 \\ -1 \\ 0 \end{bmatrix} + \lambda_3 \begin{bmatrix} -2 \\ 0 \\ 0 \end{bmatrix} + \lambda_4 \begin{bmatrix} 0 \\ 0 \\ -1 \end{bmatrix}$$

$$\leqslant \theta \begin{bmatrix} \lambda_1 + \lambda_3 + \lambda_4 \\ \lambda_2 + \lambda_3 + \lambda_4 \\ \lambda_4 \end{bmatrix}, \qquad \lambda_i \geqslant 0, \qquad i = 1,2,3,4$$

which is solved, for every $0 < \theta \leqslant \theta^* = \frac{1}{2}$, by

$$\lambda_1 = \frac{2(1-2\theta)}{\theta(1-\theta)^2}, \qquad \lambda_2 = 0, \qquad \lambda_3 = \frac{1-2\theta}{\theta(1-\theta)}, \qquad \lambda_4 = \frac{1}{1-\theta}.$$

Thus (DG.θ) establishes the optimality. (Note that two of the multipliers blow to infinity when $\theta \to 0$.)

SUGGESTED FURTHER READING

ABRAMS [75], ABRAMS and KERZNER [78], BEN-ISRAEL and BEN-TAL [76], BEN-ISRAEL, BEN-TAL and ZLOBEC [79], BEN-TAL and BEN-ISRAEL [76] and [79], BEN-TAL, BEN-ISRAEL and ZLOBEC [76], BEN-TAL and CHARNES [76], BORWEIN and WOLKOWICZ [79], CRAVEN and ZLOBEC [78], EREMIN and ASTAFIEV [76], GUIGNARD [69], MANGASARIAN [69], MASSAM [79], MAURER and ZOWE [79], MOND and ZLOBEC [79], ROBERTS and VARBERG [73], ROBINSON [76], ROCKAFELLAR [70a], STOER and WITZGALL [70], WOLKOWICZ [78 and 79], ZLOBEC and BEN-ISRAEL [79a and 79b], ZLOBEC, GARDNER, and BEN-ISRAEL [80], ZLOBEC and JACOBSON [79].

2
SOME COMPUTATIONAL METHODS

5 THE METHOD OF FEASIBLE DIRECTIONS

In this chapter we consider the convex program

(P)

$$\min f^0(x)$$

s.t.

$$f^k(x) \leqslant 0, \qquad k \in \mathscr{P} = \{1, 2, \ldots, p\}$$

$$x^T c^j + \beta_j \leqslant 0, \qquad j \in \mathscr{T} = \{1, 2, \ldots, q\}$$

$$L \leqslant x \leqslant U$$

where $f^k : R^n \to R$, $k \in \{0\} \cup \mathscr{P}$ are convex (nonlinear) differentiable functions, $c^j \in R^n$, $\beta_j \in R$, $j \in \mathscr{T}$, $L = (L_i) \in R^n$, and $U = (U_i) \in R^n$. Denote by S the set of feasible solutions of (P). It is assumed that S satisfies Slater's condition, that is,

$$\begin{cases} \exists \hat{x} \quad \text{such that} \quad L \leqslant \hat{x} \leqslant U, \qquad \hat{x}^T c^j + \beta_j \leqslant 0, \qquad j \in \mathscr{T} \text{ and} \\ \qquad f^k(\hat{x}) < 0, \qquad k \in \mathscr{P}. \end{cases} \tag{5.1}$$

The method of feasible directions is an iterative method for solving problem (P). The lth iteration starts with a feasible solution $x^l = x$ and (unless x is optimal) generates an improved solution

$$x_{\text{NEW}} = x + \alpha^* \bar{d}.$$

Here $\bar{d}$ is a feasible direction of S at x, which is also a direction of descent of f^0, that is,

$$\bar{d} \in D_{f^0}(x) \cap F(S, x).$$

The *step size* α^* is a solution of the (single variable) *constrained line search problem*

(LS)

$$\min\left\{ f^0(x + \alpha\bar{d}): \alpha \geqslant 0, \qquad x + \alpha\bar{d} \in S \right\}.$$

This problem is of independent interest and is thoroughly discussed in Section 8.

In Zoutendijk's classical method (ZOUTENDIJK [60], [76]) the following *direction generator* is used for calculating $\bar{d}$:

(Z)

$$\max \lambda$$

s.t.

$$
\begin{aligned}
\nabla f^k(x)^T d + \lambda \leqslant 0, & \qquad k \in \{0\} \cup \mathcal{P}(x) & (5.2) \\
d^T c^j \leqslant 0, & \qquad j \in \mathcal{T}(x) & (5.3) \\
0 \leqslant d_i, \quad i \in I^-(x); \quad d_i \leqslant 0, & \qquad i \in I^+(x) & (5.4) \\
\text{normalization condition } N(d). &
\end{aligned}
$$

Here $\mathcal{P}(x) = \{k \in \mathcal{P} : f^k(x) = 0\}$, $\mathcal{T}(x) = \{j \in \mathcal{T} : (x)^T c^j + \beta_j = 0\}$, $I^-(x) = \{i : x_i = L_i\}$, $I^+(x) = \{i : x_i = U_i\}$, and $N(d)$ denotes a *normalization condition*, such as

$$
\begin{aligned}
N_1: & \quad |d_i| \leqslant 1, \qquad i = 1, 2, \ldots, n \\
N_2: & \quad \sum_{i=1}^{n} |d_i| \leqslant 1
\end{aligned}
$$

or

$$N_3: \quad \sum_{i=1}^{n} d_i^2 \leqslant 1.$$

The normalization condition guarantees compactness of the feasible set of (Z). Note also that $d = 0$, $\lambda = 0$ is feasible. Hence an optimal solution exists, say $\bar{d}, \bar{\lambda}$ with $\bar{\lambda} \geqslant 0$. If $\bar{\lambda} > 0$, then $\bar{d}$ is a feasible direction of descent. On the other hand, if $\bar{\lambda} = 0$, then (and only then) is x optimal. Indeed, $\bar{\lambda} = 0$ if and only if the system

$$
\left\{
\begin{array}{l}
\nabla f^k(x)^T d < 0, \qquad k \in \{0\} \cup \mathcal{P}(x) \\
\text{with (5.3) and (5.4)}
\end{array}
\right.
\qquad (5.5)
$$

has no solution d. The inconsistency of (5.5) is a primal version of the Fritz

John condition for problem (P), which is necessary and sufficient for optimality of x [since Slater's condition (5.1) is satisfied].

A phenomenon associated with the feasible direction methods is the possibility that the sequence $\{x^l\}$ may be jammed in a "corner" of the feasible set and thus converge to a nonoptimal point. This phenomenon, known as *jamming*, is due mainly to the sudden change of the direction generator when new constraints become binding (see, for example, ZANGWILL [67], ZOUTENDIJK [76]). One way to overcome this difficulty is to consider not merely binding constraints, but also those that are "almost binding" in (Z). The binding constraints in (Z) are thus replaced by the ϵ-active index sets

$$\mathcal{P}_\epsilon(x) = \left\{ k \in \mathcal{P} : -\epsilon \leqslant f^k(x) \leqslant 0 \right\}$$

$$\mathcal{T}_\epsilon(x) = \left\{ j \in \mathcal{T} : -\epsilon \leqslant x^T c^j + \beta_j \leqslant 0 \right\}$$

$$I_\epsilon^-(x) = \{ i : L_i \leqslant x_i \leqslant L_i + \epsilon \} \tag{5.6}$$

$$I_\epsilon^+(x) = \{ i : U_i - \epsilon \leqslant x_i \leqslant U_i \}$$

for some $\epsilon > 0$, which is decreased during the iterations. We refer to this procedure as use of the $(Z)_\epsilon$ *generator*. Another *antijamming* approach (TOPKIS and VEINOTT [67]) is to consider *all* constraints in (P). The direction generator is then

$$\max \lambda$$

s.t.

$$\nabla f^0(x)^T d + \lambda \qquad\ \leqslant 0$$

$$\nabla f^k(x)^T d + \lambda + f^k(x) \leqslant 0, \qquad k \in \mathcal{P}$$

$$d^T c^j + x^T c^j + \beta_j \qquad \leqslant 0, \qquad j \in \mathcal{T}$$

$$L \leqslant d + x \leqslant U.$$

Another difficulty associated with the above-mentioned feasible direction methods is their inability to generate feasible directions that remain on the boundary of the feasible set corresponding to the nonlinear constraints. Indeed, at a nonoptimal point x, the direction $\bar{d}$ [generated by solving (Z)] satisfies, in particular,

$$\nabla f^k(x)^T \bar{d} < 0, \qquad k \in \mathcal{P}(x).$$

This means that $\bar{d}$ points into the *interior* of the set

$$C = \left\{ x : f^l(x) \leqslant 0, \qquad k \in \mathcal{P} \right\}.$$

Let us point out that every vector $d \in D_k^=(x)$, $k \in \mathcal{P}(x)$ is a direction along the boundary of C. Since typically the constancy cone $D_k^=$ is not equal to zero (except in some special cases such as strictly convex f^k) many useful directions may be lost. For example, if

$$f^k(x) = \Phi(Ax)$$

then every direction $d \in N(A)$, that is, $Ad = 0$, satisfies $f^k(x) = f^k(x + \alpha d)$ for all α, and so remains on the boundary of $\{x : f^k(x) \leqslant 0\}$. In particular, such situations occur in programs (P) with sparse nonlinear constraints.

An extreme example of what can be gained by allowing movements along the boundary is illustrated by the following program:

$$\min f^0 = x_1 + x_2$$

s.t.

$$f^1 = x_1^2 - x_1 \leqslant 0 \qquad\qquad (5.7)$$

$$f^2 = x_2^2 - x_2 \leqslant 0.$$

The feasible set is the unit square and the optimal solution is $x^* = (0,0)^T$. Starting from the initial feasible point $x^0 = (0, 1)^T$, the method of feasible directions with direction generator (Z) and normalization condition N_1, produces the infinite sequence

$$x^l = \begin{cases} \left(0, \left(\tfrac{1}{2}\right)^l\right)^T, & \text{if } l \text{ is even} \\[2mm] \left(\left(\tfrac{1}{2}\right)^l, 0\right)^T, & \text{if } l \text{ is odd} \end{cases}$$

$$l = 1, 2, \ldots$$

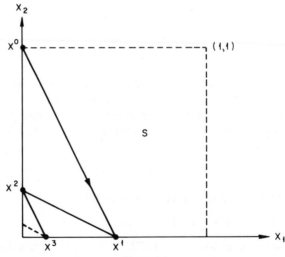

Figure 5.1

which converges to x^* (see Figure 5.1). Note, however, that had the direction (along the boundary) $d = (0, -1)^T$ been used, the optimal solution would have been found in one step.

6 MODIFIED DIRECTION-FINDING GENERATORS

The first modified direction-finding generator, which is capable of producing directions along the boundary of the feasible region, is based on the parametric characterization in Section 4. To obtain this generator requires additional work (the solution of an additional linear program at each iteration), but its use also consistently gives a direction of steeper descent than the one obtained by using the classical generator (Z).

For the sake of simplicity we consider in this chapter programs with differentiable faithfully convex functions, that is, programs (P) with

$$f^k(x) = h^k(A_k x + b^k) + x^T a^k + \alpha_k$$

where

$$h^k : R^{m_k} \to R \text{ is strictly convex}$$

$$A_k \in R^{m_k \times n}, \qquad b^k \in R^{m_k}, \qquad a^k \in R^n, \qquad \alpha_k \in R, \qquad k \in \mathcal{P}. \quad (6.1)$$

The modified $(Z)_\epsilon$-generator produces a feasible direction of descent d^L in two steps. First a vector d^Z is found by the (Z_ϵ)-generator and then the following scalars

$$\theta_k = - \frac{\nabla f^k(x)^T d^Z}{|(a^k)^T d| + \sum_{i=1}^{m_k} |A_k^i d^Z|}, \qquad k \in \mathcal{P}_\epsilon(x) \quad (6.2)$$

are calculated. [Here A_k^i is the ith row of A_k in (6.1).] Then a feasible direction of descent d^L is determined by the $(L)_\epsilon$-*generator*:

$(L)_\epsilon$

$$\min \nabla f^0(x)^T d$$

s.t.

$$\nabla f^k(x)^T d + \theta_k \left(|(a^k)^T d| + \sum_{i=1}^{m_k} |A_k^i d| \right) \leq 0, \qquad k \in \mathcal{P}_\epsilon(x)$$

$$d^T c^j \leq 0, \qquad j \in \mathcal{I}_\epsilon(x) \quad (6.3)$$

$$0 \leq d_i, \qquad i \in I_\epsilon^-(x); \qquad d_i \leq 0, \qquad i \in I_\epsilon^+(x)$$

$$N(d).$$

The following theorem is the basis underlying the use of the $(L)_\epsilon$-generator.

6.1   Theorem.   Let x be a feasible but nonoptimal solution of (P), (λ_Z, d^Z) an optimal solution of $(Z)_\epsilon$ and d^L an optimal solution of $(L)_\epsilon$. Then

    (a) d^L is a feasible direction;

    (b) d^L is of steeper descent than d^Z, that is,

$$\nabla f^0(x)^T d^L \leqslant \nabla f^0(x)^T d^Z. \tag{6.4}$$

Proof.

    (a)   Since x is not optimal, $\lambda_Z > 0$. So, by (5.2),

$$\nabla f^k(x)^T d^Z < 0, \qquad k \in \mathcal{P}_\epsilon(x). \tag{6.5}$$

    Further, by the faithful convexity,

$$\nabla f^k(x)^T d^Z = \left[\nabla h^k\left(A_k x + b^k\right)\right]^T\left(A_k d^Z\right) + \left(a^k\right)^T d^Z, \qquad k \in \mathcal{P}$$

    which, together with (6.5), shows that the denominator in the definition of θ_k is not zero. Hence θ_k is well defined and, moreover, positive. Examining the constraints (6.3), it is concluded that either $\nabla f^k(x)^T d^L < 0$ [i.e., $d^L \in D_k^<(x)$] or $A_k d^L = 0$, $(a^k)^T d^L = 0$ [i.e., $d^L \in D_k^=(x)$ (see Lemma 1.6)], which shows that d^L is a feasible direction.

    (b)   By the definition of θ_k, and (6.3), it is easily verified that $d = d^Z$ is a feasible solution of $(L)_\epsilon$ but not necessarily optimal. Hence (6.4) follows.   ∎

Let us illustrate the $(L)_\epsilon$-generator by reconsidering problem (5.7). Note that f^1 and f^2 are faithfully convex with $h^1(t_1, t_2) = h^2(t_1, t_2) = t_1^2 + t_2^2$,

$$A_1 = \begin{bmatrix} 1 & 0 \\ 0 & 0 \end{bmatrix}, \qquad A_2 = \begin{bmatrix} 0 & 0 \\ 0 & 1 \end{bmatrix},$$

$$a^1 = \begin{bmatrix} -1 \\ 0 \end{bmatrix}, \qquad a^2 = \begin{bmatrix} 0 \\ -1 \end{bmatrix}, \qquad b^1 = b^2 = \begin{bmatrix} 0 \\ 0 \end{bmatrix}, \qquad \alpha_1 = \alpha_2 = 0.$$

At $x^0 = (0, 1)^T$ one finds $\mathcal{P}(x^0) = \{1, 2\}$. The $(Z)_\epsilon$-generator ($\epsilon > 0$ arbitrary and normalization N_1) produces $d^Z = (\frac{1}{4}, -\frac{1}{2})^T$ (see Figure 5.1). The θ_k's are here

$$\theta_1 = \theta_2 = \tfrac{1}{2}$$

and the solution of $(L)_\epsilon$ is $d^L = (-1, 0)^T$. This is indeed a direction on the boundary of the feasible region: a direction that, in fact, points toward the optimal solution. The feasible direction method, with the $(L)_\epsilon$-generator, thus terminates in one iteration.

A different direction-generator is based on the multi-Ω primal characterization (see Table 3.1). For a given subset $\Omega \subset \mathcal{P}(x)$, let $\lambda = \lambda(\Omega)$, $d = d(\Omega)$ be an optimal solution of

(Z, Ω)

$$\max \lambda$$

s.t.

$$\nabla f^0(x)^T d + \lambda \leqslant 0$$
$$\nabla f^k(x)^T d + \lambda \leqslant 0, \qquad k \in \mathcal{P}(x) \backslash \Omega$$
$$A_k d = 0, \qquad (a^k)^T d = 0, \qquad k \in \Omega$$
$$(5.3), (5.4) \text{ and } N(d).$$

Then for a feasible but not optimal x, $\lambda(\Omega) > 0$ and $d(\Omega)$ is a feasible direction of descent. Note that (Z, Ω), for $\Omega = \varnothing$, is exactly the direction generator (Z). If $\Omega \neq \varnothing$, then the constraints $\{ f^k : k \in \Omega \}$ remain binding, and hence $d(\Omega)$ points along the boundary of the feasible set. In the kth order (Z, Ω)-generator, k subsets Ω of $\mathcal{P}(x)$ are considered, say $\Omega_1, \ldots, \Omega_k$. The direction chosen for descent is $d = d(\Omega^*)$, where

$$\lambda(\Omega^*) = \max_{j=1,\ldots,k} \lambda(\Omega_j) > 0,$$

which implies

$$\nabla f^0(x)^T d(\Omega^*) \leqslant \nabla f^0(x)^T d(\Omega_j) \leqslant 0, \qquad j = 1, \ldots, k;$$

thus, among k usable directions, the chosen one has steepest descent. In practice (see Ex. 6.4) one uses two subsets, one of which is always $\Omega_1 = \varnothing$. The second is Ω_2, for which $\lambda(\Omega_2) > 0$ (if such subset exists). The search for Ω_2 starts with subsets of cardinality 1 and continues with increasing order of cardinality.

In problem (5.7), $\Omega_2 = \{1\}$. Since

$$1 = \lambda(\{1\}) > \lambda(\varnothing) = \tfrac{1}{2}$$

the direction $d(\{1\}) = (0, -1)^T$ is the one produced here by the (Z, Ω)-generator. As is mentioned earlier, this direction points toward the optimum.

When Slater's condition is not satisfied for problem (P), then $\lambda(\varnothing) = 0$ (see Ex. 6.1). However, a method based on a (Z, Ω)-generator is still applicable.

Exercises and Examples

6.1   A computational test for the validity of Slater's condition.   Let $\{f^k : k \in \mathcal{P}\}$ be a differentiable convex function and let x be a fixed but arbitrary feasible solution of problem (P). Then Slater's condition holds for (P) if and only if the optimal value λ^* of the linear program

$$\max \lambda$$

s.t.

$$\nabla f^k(x)^T d + \lambda \leqslant 0, \qquad k \in \mathcal{P}(x)$$

$$|d_i| \leqslant 1, \qquad i = 1, 2, \ldots, n$$

is positive.

Proof.   The optimal value λ^* is positive if and only if the linear program (without the normalization condition) is unbounded. This is the case if and only if dual is not feasible, that is, the system

$$\sum_{k \in \mathcal{P}(x)} \lambda_k \nabla f^k(x) = 0$$

$$\lambda_k \geqslant 0, \qquad k \in \mathcal{P}(x), \qquad \sum_{k \in \mathcal{P}(x)} \lambda_k = 1$$

is inconsistent. The latter means that x is a regular point. That is equivalent to the validity of Slater's condition (see Ex. 3.12).   ∎

THE (Z, Ω)-GENERATOR FOR PROBLEMS WITH STRICTLY CONVEX RESTRICTIONS

For problems with strictly convex restrictions (see Ex. 1.9) there is a simple way of detecting subsets Ω for which the optimal value of problem (Z, Ω) is $\lambda(\Omega) = 0$. This procedure is described and implemented in Exs. 6.2–6.5.

6.2   Consider the *elimination condition* for a proper subset $\Omega \subset \mathcal{P}(x^*)$:

(E)

$$\begin{cases} \text{there exists index } k \in \{0\} \cup [\mathcal{P}(x^*)\backslash\Omega] \text{ such that} \\ \left\{i : \dfrac{\partial}{\partial x_i} f^k(x^*) \neq 0\right\} \subset \bigcup_{j \in \Omega} [j]. \end{cases}$$

If (E) holds for a subset Ω, then $\lambda(\Omega) = 0$. Moreover, if (E) is realized by $k = k_0$, then $\lambda(\overline{\Omega}) = 0$ for every subset $\overline{\Omega}$ such that

$$k_0 \notin \overline{\Omega} \supseteq \Omega. \tag{6.6}$$

Proof. The first statement is proved as in Ex. 3.14. The second follows trivially by (6.6). ∎

6.3 Example for Ex. 6.2. Consider the problem

$$\min f^0 = \quad x_2 \quad + e^{-x_4}$$

s.t.

$$f^1 = e^{x_1} + e^{x_2} + e^{-x_3} + e^{x_4} - 4 \leqslant 0$$

$$f^2 = e^{x_1} + (1 - x_3)^2 + e^{-x_4} - 3 \leqslant 0$$

$$f^3 = x_1^2 + e^{-x_2} + (1 + x_4)^2 - 2 \leqslant 0$$

$$f^4 = e^{-x_1} + e^{x_2} + 2e^{x_3} \quad - 5 \leqslant 0$$

$$f^5 = x_1^2 + e^{x_2} + e^{x_3} + e^{-x_4} - 3 \leqslant 0$$

and the feasible point $x = 0$, at which $\mathcal{P}(x) = \{1, 2, 3, 5\}$. To identify the subsets that are eliminated by (E), we associate with the above program, at $x = 0$, its *incidence matrix* $A = (a_{kj})$, $k \in \{0\} \cup \mathcal{P}(x)$, $j = 1, 2, \ldots, n$, where

$$a_{kj} = \begin{cases} 0 & \text{if} \quad \dfrac{\partial}{\partial x_j} f^k(x) = 0 \\ 1 & \text{otherwise.} \end{cases}$$

Here

	j			
k	1	2	3	4
0	0	1	0	1
1	1	1	1	1
2	1	0	1	1
3	0	1	0	1
5	0	1	1	1

Clearly $\Omega = \{1\}$ is eliminated, since [1] contains all variables. Similarly, $\{2, 3\}$ and $\{2, 5\}$ are eliminated. Thus all 11 subsets, containing at least one of the above, are also eliminated. For the subsets $\{3\}$, $\{5\}$, and $\{3, 5\}$, (E) holds for $k = 0$. Therefore, out of the $2^4 - 1 = 15$ proper subsets of $\mathcal{P}(x)$, all are eliminated but $\Omega = \{2\}$.

6.4 A description of the (Z, Ω)-generator of order $k = 2$ (discussed in the text above) is now given. The method involves use of the elimination condition and hence is termed MELP (abbreviation for "method of elimination of linear programs").

Step 1   Solve $(Z, \emptyset)$ to determine its optimal solution $[\lambda(\emptyset), d(\emptyset)]$. If $\mathscr{P}(x) = \emptyset$, set $d = d(\emptyset)$; stop.

Step 2   Among all subsets of $\mathscr{P}(x)$ determine a noneliminated subset Ω_2 of minimal cardinality for which $\lambda(\Omega_2) > 0$. [If no such subset exists, set $d = d(\emptyset)$.]

Step 3   Solve (Z, Ω_2) to determine its optimal solution $[\lambda(\Omega_2), d(\Omega_2)]$.

Step 4   Determine the *set* Ω^* by

$$\Omega^* = \begin{cases} \emptyset, & \text{if } \lambda(\emptyset) \geqslant \lambda(\Omega_2) \\ \Omega_2 & \text{otherwise} \end{cases}$$

and set $d = d(\Omega^*)$; stop.

6.5   An implementation of MELP is carried out on the following problem.

$$\min f^0(x) = x_1 - x_2 + (x_3 - 1)^2 + (x_4 - 2)^2 + (x_5 - 3)^2$$

s.t.

$$
\begin{aligned}
f^1(x) &= e^{x_1} + x_2^2 & & & -2 &\leqslant 0 \\
f^2(x) &= x_1^2 + x_2^2 + e^{-x_3} & & & -2 &\leqslant 0 \\
f^3(x) &= x_1 & & +x_4^2 + x_5^2 & -2 &\leqslant 0 \\
f^4(x) &= x_1^2 + x_2^2 - 4x_2 & & & -1 &\leqslant 0. \\
f^5(x) &= (x_1 - 1)^2 + x_2^2 & & & -2 &\leqslant 0 \\
f^6(x) &= x_1 & & +e^{-x_4} & -2 &\leqslant 0 \\
f^7(x) &= \quad\quad x_2 & & +e^{-x_5} & -2 &\leqslant 0.
\end{aligned}
$$

The starting point is $x^0 = (0, 0, 0, 0, 0)^T$, for which none of the constraints is binding. Therefore the direction $d^Z = d(\emptyset) = (-1, 1, 1, 1, 1)^T$ is used. The next feasible point is $x^1 = (-0.366, 0.366, 0.366, 0.366, 0.366)^T$, at which $\mathscr{P}(x^1) = \{5\}$. Since

$$9.804 = \lambda(\{5\}) > \lambda(\emptyset) = 3.464$$

the pivot set is $\Omega^* = \Omega_2 = \{5\}$ and the direction used is $d(\{5\}) = (0, 0, 1, 1, 1)^T$, which yields $x^2 = (-0.366, 0.366, 1.088, 1.088, 1.088)^T$, at which $f^0(x^2) = 3.765$. The corresponding value, when using $d^Z = d(\emptyset) = (1, -1, 0, 0.060, 1)^T$, is $f^0(x^2) = 7.581$.

CONVERTING A PROBLEM INTO ONE FOR WHICH SLATER'S CONDITION HOLDS

Every program $(P): \min\{f^0(x): f^k(x) \leqslant 0, \, k \in \mathscr{P}\}$ can be converted into one that satisfies Slater's condition, for example,

$$\min f^0(x) + M x_{n+1}$$

s.t.

$$f^k(x) - x_{n+1} \leqslant 0 \quad k \in \mathscr{P} \tag{6.7}$$
$$- x_{n+1} \leqslant 0.$$

Indeed, $x^* \in R^n$ is an optimal solution of (P) if and only if, for M large enough, $(x^*, 0)^T \in R^{n+1}$ is an optimal solution of (6.7). Usually it suffices to associate x_{n+1} with only a small number of constraints.

6.6 Consider

$$\min f^0(x) = x_1 + x_2 + x_3$$

s.t.

$$f^1(x) = x_1^2 + x_2^2 \qquad\qquad -2 \leqslant 0$$
$$f^2(x) = (x_1 - 2)^2 + (x_2 - 2)^2 - 2 \leqslant 0$$
$$f^3(x) = \qquad\qquad e^{-x_3} \quad -1 \leqslant 0.$$

The feasible set is $\{(1, 1, x_3)^T : x_3 \geqslant 0\}$ and it does not satisfy Slater's condition. Starting from any feasible solution, the MELP obtains the optimal solution $x^* = (1, 1, 0)^T, f^0(x^*) = 2$, in *one iteration*. Now we demonstrate what can happen when (6.8) is converted to a program that does satisfy Slater's condition. Such a program, with the penalty $M = 100$, is

$$\min x_1 + x_2 + x_3 \qquad + 100 x_4$$

s.t.

$$x_1^2 + x_2^2 \qquad\qquad -2 \leqslant 0$$
$$(x_1 - 2)^2 + (x_2 - 2)^2 \qquad - x_4 - 2 \leqslant 0$$
$$e^{-x_3} \qquad -1 \leqslant 0$$
$$- x_4 \qquad \leqslant 0.$$

Starting from the initial point $x^0 = (1, 1, 1, 0)^T$, the iterations are calculated by the (Z)-generator and are summarized in Table 6.1. Convergence is obviously very slow (regardless of the choice of x^0).

TABLE 6.1

Iteration	x_1	x_2	x_3	x_4	Value of the objective function
0	1	1	1	1	103
1	0.776	0.776	1	0.99552	102.104
2	1	1	1	0.98024	101.024
3	0.780	0.780	1	0.97585	100.145
4	1	1	1	0.96087	99.087
5	0.784	0.784	1	0.95656	98.224
6	1	1	1	0.94189	97.189
7	0.788	0.788	1	0.93765	96.341
8	1	1	1	0.92328	95.328

7 THE PARAMETRIC FEASIBLE-DIRECTION ALGORITHM

This section presents a *parametric feasible-direction* (PFD) algorithm, in which use is made of the $(L)_\epsilon$-generator, for solving problem (P). The description of the algorithm includes a specific antijamming procedure, a normalization condition, and a stopping rule. The procedure for computing the step size is given in the next section. A method for calculating an initial feasible solution x^0 is described in Ex. 7.1.

7.1 PFD Algorithm

Initialization. Specify $\epsilon_0 > 0$ ("ϵ-activity parameter") and $0 < \epsilon' < \epsilon_0$ ("stopping rule parameter"). Choose a scalar δ such that $n - 1 < \delta < n$. Find an initial feasible solution x^0. Set $k = 0$.

Step 1 Set $x = x^k$, $\epsilon = \epsilon_k$. Solve $(Z)_\epsilon$ with the normalization condition

N_4:

$$|d_i| \leqslant 1, \qquad i = 1, 2, \ldots, n; \qquad \sum_{i=1}^{n} |d_i| \leqslant \delta.$$

Let λ_Z, d^Z denote the solution. If $\lambda_Z = 0$, stop (x is optimal); otherwise set $\lambda = \lambda_Z$, $d = d^Z$.

Step 2 If $\mathcal{P}_\epsilon(x) = \varnothing$, go to step 4. Otherwise continue.

Step 3 Calculate θ_k by formula (6.2); solve $(L)_\epsilon$ with the normalization N_4. Denote the solution by d and set $\lambda = \nabla f^0(x)^T d$.

Step 4 If $\lambda > \epsilon'$, solve the line search problem (LS) to determine the step size $\bar{\alpha}$. Set $x^{k+1} = x + \bar{\alpha}d$ and continue. Otherwise set $x^{k+1} = x$ and go to Step 6.

Step 5 If $\epsilon > \epsilon'$, set $\epsilon_{k+1} = \frac{1}{2}\epsilon_{k-1}$, otherwise set $\epsilon_{k+1} = \min(\epsilon, \lambda)$. Return to step 1.

Step 6 If $\epsilon \leqslant \epsilon'$, stop; x^{k+1} is optimal. Otherwise set $\epsilon_{k+1} = \min(\epsilon, \lambda)$ and return to step 1.

Remarks

7.2 The normalization condition N_4. Being linear, the normalization condition $N_1 : -1 \leqslant d_i \leqslant 1$, $i = 1, 2, \ldots, n$ is computationally convenient, but it may cause overall slow convergence. On the other hand, the normalization condition $N_3 : \sum_{i=1}^{n} d_i^2 \leqslant 1$ is frequently more effective, but nonlinear (and hence requires more work). The normalization condition N_4, used in the PFD algorithm, is a compromise between the two (for $n = 2$ and $\delta = \sqrt{2}$ it is depicted in Figure 7.1) and still retains linearity.

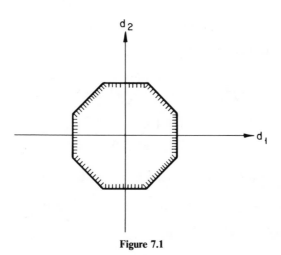

Figure 7.1

7.3 Solving $(L)_\epsilon$. Because of the "absolute values" in (6.3) and the normalization N_4, the program $(L)_\epsilon$ is not linear. An associated linear program is derived as follows.

Denote

$$\hat{A}_k = \begin{bmatrix} A_k \\ (a^k)^T \end{bmatrix}$$

and its ith row by $\hat{A}_k^i$. Observe that (6.3) is equivalent to

$$\nabla f^k(x)^T d + \theta_k \sum_{i=1}^{m_k+1} |y_i^k| \leq 0 \qquad (7.1)$$

$$\hat{A}_k d + y^k = 0, \qquad k \in \mathcal{P}_\epsilon(x). \qquad (7.2)$$

Next we introduce new variables

$$u_i^k = \tfrac{1}{2}(|y_i^k| + y_i^k)$$
$$v_i^k = \tfrac{1}{2}(|y_i^k| - y_i^k), \qquad i = 1, 2, \ldots, m_k + 1.$$

Note that

$$u_i^k \geq 0, \qquad v_i^k \geq 0, \qquad u_i^k + v_i^k = |y_i^k|, \qquad u_i^k - v_i^k = y_i^k,$$
$$i = 1, 2, \ldots, m_k + 1$$

and that, by (7.2),

$$v^k = \hat{A}_k d + u^k \geq 0.$$

So (7.1)–(7.2) can be replaced by

$$\begin{cases} \nabla f^k(x)^T d + \theta_k \sum_{i=1}^{m_k+1} (2u_i^k + \hat{A}_k^i d) \leq 0 \\ \hat{A}_k d + u^k \geq 0 \\ u^k \geq 0, \qquad k \in \mathcal{P}_\epsilon(x). \end{cases} \qquad (7.3)$$

The absolute values in the normalization condition are expressed by

$$d_i = d_i^+ - d_i^-, \qquad |d_i| = d_i^+ + d_i^-, \qquad d_i^+ \geq 0, \qquad d_i^- \geq 0,$$
$$d_i^+ d_i^- = 0, \qquad i = 1, 2, \ldots, n. \qquad (7.4)$$

The program $(L)_\epsilon$ is thus replaced by

(LP)

$$\min \nabla f^0(x)^T (d^+ - d^-)$$

s.t.

$$\nabla f^k(x)^T (d^+ - d^-) + \theta_k \sum_{i=1}^{m_k+1} \left[2u_i^k + \hat{A}_k^i (d^+ - d^-) \right] \leqslant 0$$

$$\hat{A}_k (d^+ - d^-) + u^k \geqslant 0$$

$$u^k \geqslant 0, \qquad k \in \mathscr{P}_\epsilon(x)$$

$$(d^+ - d^-)^T c^j \leqslant 0, \qquad j \in \mathscr{T}_\epsilon(x)$$

$$0 \leqslant d_i^+ - d_i^-, \qquad i \in I_\epsilon^-(x); \qquad d_i^+ - d_i^- \leqslant 0, \qquad i \in I_\epsilon^+(x)$$

$$d_i^+ + d_i^- \leqslant 1, \qquad d_i^+ \geqslant 0, \qquad d_i^- \geqslant 0, \qquad i = 1, 2, \ldots, n$$

$$\sum_{i=1}^n (d_i^+ + d_i^+) \leqslant \delta$$

$$d_i^+ d_i^- = 0, \qquad i = 1, 2, \ldots, n \tag{7.5}$$

with unknowns $d^+ \in R^n$, $d^- \in R^n$, $u^k \in R^{m_k+1}$, $k \in \mathscr{P}_\epsilon(x)$. Program (LP) is solved by the simplex algorithm *ignoring the orthogonality condition* (7.5). Indeed, if d^+, d^-, u^k, $k \in \mathscr{P}_\epsilon(x)$ solves (LP) without (7.5), then the same u^k, $k \in \mathscr{P}_\epsilon(x)$ and

$$\hat{d}_i^+ = \max(0, d_i^+ - d_i^-), \qquad \hat{d}_i^- = \max(0, d_i^- - d_i^+)$$

is also an optimal solution, since

$$\hat{d}_i^+ \hat{d}_i^- = 0$$

$$\hat{d}_i^+ - \hat{d}_i^- = d_i^+ - d_i^-$$

and

$$\hat{d}_i^+ + \hat{d}_i^- = |d_i^+ - d_i^-| \leqslant d_i^+ + d_i^-, \qquad i = 1, 2, \ldots, n.$$

The linear program (LP) is of larger size that $(Z)_\epsilon$, but its coefficient matrix is sparse and structured. It has the form

$$\begin{bmatrix} M & -M & U \\ I & -I & 0 \\ I & I & 0 \end{bmatrix}$$

where the submatrix U corresponding to the vector variables u^k, $k \in \mathscr{P}(x)$

is itself sparse and structured

$$U = \begin{bmatrix} c^1 & & & & & & 0 \\ & c^2 & & & & & \\ & & \ddots & & & & \\ 0 & & & & c^p & & \\ \cdots & & \cdots & & \cdots & & \cdots \\ & & & I & & & \end{bmatrix}$$

where c^k is an $m_k + 1$ row vector.

7.4   The antijamming procedure used in the algorithm is the one suggested by MEYER [75].

Exercises and Examples

7.1   Finding a feasible point of program (P).   A point $\bar{x}$ that satisfies the linear constraints of (P) is found by, say, phase I of the simplex method. If also $f^k(\bar{x}) \leqslant 0$ for all $k \in \mathcal{P}$, then $\bar{x}$ is the sought feasible solution of (P). Otherwise, one chooses numbers η_k, such that

$$\eta_k = \begin{cases} 0 & \text{if } f^k(\bar{x}) \leqslant 0 \\ 1 & \text{if } f^k(\bar{x}) > 0, \quad k \in \mathcal{P} \end{cases}$$

and solves

$$\min \lambda$$

s.t.

$$\begin{aligned} f^k(x) - \eta_k \lambda &\leqslant 0, \quad k \in \mathcal{P} \\ x^T c^j + \beta_j &\leqslant 0, \quad j \in \mathcal{J} \\ L \leqslant x &\leqslant U. \end{aligned} \tag{7.6}$$

Program (7.6) is solved by a feasible direction method with the initial feasible solution

$$\tilde{x} = \bar{x}, \qquad \tilde{\lambda} = \max\{ f^k(\bar{x}) : \eta_k = 1 \}.$$

Since Slater's condition holds for program (P), the optimal value of (7.6) is negative. Hence, in finitely many iterations, a feasible solution (λ^0, x^0) with $\lambda_0 \leqslant 0$ is reached and x^0 is a feasible solution of (P).

7.2   The following example is taken from SANDBLOM [73]. It shows how the choice of the normalization condition can drastically affect the convergence of Algorithm 7.1.

$$\min(x_1 - 7)^2 - x_2 + e^{-x_3} + x_3 + x_4^2 + e^{x_5} + (x_6 + 3)^4 + (x_7 - 4)^2$$
$$+ e^{-x_8} + 12x_8 + e^{x_9} + (x_9 + x_{10})^2 + (x_{11} - x_{12})^4 + e^{x_{13}} + e^{-x_{14}}$$
$$+ (x_{15} + x_{16})^2 + e^{x_{17}} - 2x_{17} + x_{18}^2$$

s.t.

x_1^2				$\leqslant$ 1.5
$x_1^2 + e^{x_2}$				$\leqslant$ 2
$-2x_1 + (x_3 - 1)^2$				$\leqslant$ 4
$e^{x_1} + x_1$	$+ x_4$			$\leqslant$ 5
$(x_1 + 2)^2$	$+ e^{-x_5} - x_5$			$\leqslant$ 11
$e^{-x_1} + x_1^2$	$+ x_7^2$			$\leqslant$ 15
$(x_1 - 3)^2$	$- x_8$			$\leqslant$ 12
$-3x_1$	$+ (x_9 - x_{10})^2$			$\leqslant$ 2
$x_1^4 + x_1$		$+ x_{11} + e^{-x_{12}}$		$\leqslant$ 12
$e^{x_1} - 10x_1$		$+ x_{13}^2 - 5x_{14}$		$\leqslant$ 20
$x_1^4 - 5x_1$		$+ e^{-x_{15}} + (x_{15} - x_{16})^2$		$\leqslant$ 35
$(x_1 + 4)^4$			$+ e^{x_{17}} - 10x_{18}$	$\leqslant$ 300
$7x_1$	$- 2x_6$			$\leqslant$ 10

The problem was solved by the PFD algorithm starting with the feasible solution

$$x_1 = 1, \quad x_2 = 0, \quad x_3 = 2, \quad x_4 = 1, \quad\quad x_5 = 0, \quad x_6 = -1.5$$
$$x_7 = 2, \quad x_8 = -8, \quad x_9 = 1, \quad x_{10} = 0, \quad\quad x_{11} = 9, \quad x_{12} = 0$$
$$x_{13} = 5, \quad x_{14} = 1, \quad x_{15} = 0, \quad x_{16} = 6.16441, \quad x_{17} = 0, \quad x_{18} = 32.6001$$

using two different normalizations N_1 and N_4. The results are compared in Table 7.1.

TABLE 7.1

The objective function value $f^0(x^\ell)$

ℓ	N_1	N_4
0	10,749.40	10,749.40
1	7,461.48	7,711.90
5	2,066.90	218.40
10	797.25	59.71
30	322.98	30.89
50	139.53	30.60
100	77.39	

In some applications the number of iterations is indeed more important than the total CPU time spent on computing the iterations themselves. This is particularly true in optimal structure design where the number of times one has to analyze the structure, which could be a time-consuming process, is proportional to the number of iterations. (See, for example, VANDERPLAATS and MOSES [73] and SHEU [75].)

7.3 A numerical example for the PFD method.

The problem

$$\min x_1 - x_2 + (x_3 - 1)^2 + (x_4 - 2)^2 + (x_5 - 2)^2$$

s.t.

$$
\begin{aligned}
e^{x_1} + x_2^2 &\leqslant 2 \\
x_1^2 + x_2^2 + e^{-x_3} &\leqslant 2 \\
x_1 \qquad\qquad + x_4^2 + x_5^2 &\leqslant 2 \\
(x_1 - 1)^2 \qquad + x_2^2 &\leqslant 2 \\
x_1 \qquad + e^{-x_4} &\leqslant 2 \\
x_2 \qquad\qquad + e^{-x_5} &\leqslant 2 \\
- x_2 &\leqslant \ln 2
\end{aligned}
$$

was solved by the PFD method, starting from the feasible point $x^0 = 0$, with a six decimal place accuracy ($\epsilon' = 10^{-6}$). The final solution, obtained in 1.38 seconds (CPU time), is $x^{35} = (-0.249425, 0.662524, 1.00000, 1.06052, 1.06052)^T$, see Table 7.2.

When the same problem was solved starting from the nonfeasible point $x^0 = (5, 5, 5, 5, 5)^T$ and using the procedure outlined in Ex. 7.1, the same solution was obtained in double the time.

TABLE 7.2

Iteration	Objective function value	Step size
0	9	1.48318
1	1.94112	0.166564
5	0.910488	0.210319
10	0.857503	$4.41644 \cdot 10^{-2}$
20	0.853591	$1.13018 \cdot 10^{-4}$
30	0.853280	$1.42157 \cdot 10^{-6}$
35	0.853280	

8 SOLVING CONSTRAINED LINE SEARCH PROBLEMS

Given a feasible direction of descent $\bar{d}$, generated by the PFD method, we now have the problem of determining the step size α^*, that is, an optimal solution of the constrained line search problem

(LS)

$$\min f^0(x + \alpha\bar{d})$$

s.t.

$$f^k(x + \alpha\bar{d}) \leqslant 0, \qquad k \in \mathcal{P} \tag{8.1}$$

$$(x + \alpha\bar{d})^T c^j + \beta_j \leqslant 0, \qquad j \in \mathcal{J} \tag{8.2}$$

$$L_i \leqslant x_i + \alpha\bar{d}_i \leqslant U_i, \qquad i = 1, 2, \ldots, n \tag{8.3}$$

$$\alpha \geqslant 0.$$

The constraints (8.2) and (8.3) reduce to $\alpha \leqslant \bar{\alpha}$, where

$$\bar{\alpha} = \min(\alpha_1, \alpha_2, \alpha_3)$$

$$\alpha_1 = \min_{j \notin \mathcal{J}(x)} \left\{ \frac{-\beta^j - x^T c^j}{\bar{d}^T c^j} : \bar{d}^T c^j > 0 \right\}$$

$$\alpha_2 = \min_{i \notin I^+(x)} \left\{ \frac{U_i - x_i}{\bar{d}_i} : \bar{d}_i > 0 \right\}$$

$$\alpha_3 = \min_{i \notin I^-(x)} \left\{ \frac{L_i - x_i}{\bar{d}_i} : \bar{d}_i < 0 \right\}.$$

(We use the convention that the minimum over an empty set is $+\infty$.) Some of the constraints in (8.1) may be superfluous. Indeed, if $\mathcal{P}(x, \bar{d})$ denotes those indices of $\mathcal{P}_\epsilon(x)$ that are binding at the optimal solution $\bar{d}$ of $(L)_\epsilon$, then for every $k \in \mathcal{P}(x, \bar{d})$ the corresponding constraint in (8.1) is redundant. The reason is that for such k's, $\bar{d} \in D_k^=(x)$, implying

$$f^k(x + \alpha\bar{d}) = f^k(x) \leqslant 0 \qquad \forall \ \alpha \in R.$$

Combining the above, we see that (LS) becomes

$$\min g_0(\alpha)$$

s.t.

$$g_k(\alpha) \leqslant 0, \qquad k \in \mathcal{P} \backslash \mathcal{P}(x, \bar{d})$$

$$0 \leqslant \alpha \leqslant \bar{\alpha},$$

where

$$g_k(\alpha) \underset{=}{\Delta} f^k(x + \alpha \bar{d}).$$

Since $\bar{d}$ is a feasible direction of descent:

$$g_0'(0) < 0; \qquad g_k'(0) < 0, \qquad k \in \mathscr{P}_\epsilon(x) \backslash \mathscr{P}(x, \bar{d}).$$

Also

$$g_k(0) < 0, \qquad k \in \mathscr{P} \qquad \text{but} \qquad k \notin \mathscr{P}_\epsilon(x) \backslash \mathscr{P}(x, \bar{d}).$$

Hence, by the convexity of g_k's, each of the functions

$$g_0'(\alpha), \qquad g_k(\alpha), \qquad k \in \mathscr{K} \underset{=}{\Delta} \mathscr{P} \backslash \mathscr{P}(x, \bar{d})$$

has at most one positive root. Moreover, the smallest of these roots in the interval $(0, \bar{\alpha}]$ is the sought step size α^*. If there are no roots in $(0, \bar{\alpha}]$, then $\alpha^* = \bar{\alpha}$ (see Figure 8.1).

The problem of computing the step size is actually part of a more general problem, namely, that of computing the smallest of the roots of a collection of functions in a given interval. The rest of the section is devoted to this problem, subsequently referred to as the MINIROOT problem. Without loss of generality the interval [0, 1] is considered.

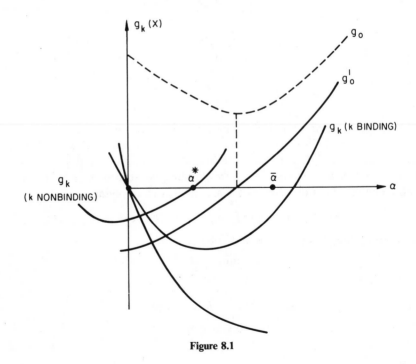

Figure 8.1

A function $\varphi_k : R_+ \to R$ is called a *switching function* on the interval $[0, 1]$ if $\varphi_k(0) < 0$ and

> *either* there exists a unique $\alpha_k \in (0, 1]$ such that
>
> $\varphi_k(\alpha) < 0$ for $\alpha \in (0, \alpha_k)$ and $\varphi_k(\alpha) > 0$ for $\alpha \in (\alpha_k, 1]$,
>
> *or* $\varphi_k(\alpha) < 0$ for $\alpha \in (0, 1]$.

We refer to α_k as the *root* of φ_k.

Problem MINIROOT. Given a collection of switching functions φ_k, $k \in \mathcal{Q} \underset{=}{\Delta} \{1, 2, \ldots, q\}$, find the minimal root $\alpha^* = \min\{\alpha_k : k \in \mathcal{Q}\}$ or assert that no root exists in the interval $(0, 1]$.

The method of solution is derived first for two functions ($q = 2$). So we consider the problem of finding the minimal root α^* for two functions φ_1 and φ_2 using N sequential "observations." Each observation consists of evaluating one of the functions at some point, determined by the preceding observations. To illustrate this dependence, let, say, φ_1 be evaluated at some $\alpha \in (0, 1)$. If $\varphi_1(\alpha) > 0$, then its root $\alpha_1 \in (0, \alpha)$, and there is no need to consider the observation points in $(\alpha, 1]$ (for either function) at later iterations. Thus before the observation the *relevant interval*, that is, the interval that may contain α^*, is $(0, 1]$ for both functions; after the observation it is $(0, \alpha)$ for both functions. On the other hand, if $\varphi_1(\alpha) \leqslant 0$, then the resulting relevant interval for φ_1 is $[\alpha, 1]$ and for φ_2 it is still $(0, 1]$. Observe that regardless of the outcome [sign of $\varphi_1(\alpha)$] of the observation:

(P1)   "the relevant intervals have common right-hand sides."

In general, if at any stage the relevant intervals are, say, $(\beta_1, \xi]$ and $(\beta_2, \xi]$, then observing φ_1 at $\alpha \in (\beta_1, \xi)$ generates the relevant intervals:

(α, ξ) and (β_2, ξ) if $\varphi_1(\alpha) \leqslant 0$

(β_1, α) and (β_2, α) if $\varphi_1(\alpha) > 0$ and $\alpha > \beta_2$

(β_1, α) if $\varphi_1(\alpha) > 0$ and $\alpha \leqslant \beta_2$.

(Note that the last-mentioned case reveals that α^* cannot be the root of φ_2, and hence φ_2 can be deleted.) The above induction argument proves that (P1) is indeed valid.

At a given stage ℓ, the point α^* is, for certain, in the bigger of the two relevant intervals, whose length I_ℓ thus measures the error of locating α^*. Our aim is to minimize the error by deciding which function φ_i to evaluate and at what point α. Each decision results in one of two possible pairs of

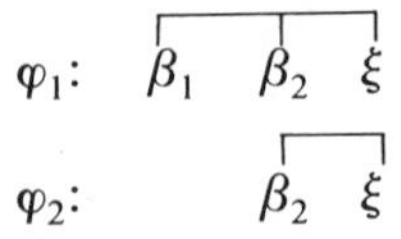

φ_1: $\overbrace{\beta_1 \quad \beta_2}^{} \quad \xi$

φ_2: $\qquad \overbrace{\beta_2}^{} \quad \xi$

Figure 8.2

relevant intervals, depending on the sign of $\varphi_i(\alpha)$. Let $I_{\ell+1}^{+} = I_{\ell+1}^{+}(\alpha, i)$ and $I_{\ell+1}^{-} = I_{\ell+1}^{-}(\alpha, i)$ be the lengths of the maximal members of each pair, corresponding to the positive and negative signs. Thus the best *a priori* error estimate is

$$\max\{ I_{\ell+1}^{+}(\alpha, i), I_{\ell+1}^{-}(\alpha, i) \}$$

and we want to find $i_{\ell}^{*} \in \{1, 2\}$ and α_{ℓ}^{*} that minimize this function.

To resolve the question of which function to observe first, consider the general situation depicted in Figure 8.2. It is assumed, without loss of generality, that the relevant interval of φ_1 is the bigger one, that is,

$$\beta_1 \leqslant \beta_2.$$

The outcome of measuring φ_1 first is summarized in Table 8.1, while Table 8.2 summarizes the outcome of measuring φ_2 first. Comparing the two tables, we deduce that for positive outcomes it is immaterial which function is observed first (the same error $\alpha - \beta_1$ is obtained). But for negative outcomes it is advantageous to measure φ_1 first, since:

$$\xi - \alpha \leqslant \xi - \beta_1 \qquad \text{for} \quad \beta_1 \leqslant \alpha \leqslant \beta_2$$

and

$$\xi - \beta_2 \leqslant \xi - \beta_1.$$

Thus we have proved:

(P2)   "at every stage it is optimal to observe the function having the largest relevant interval."

TABLE 8.1

	$I_{\ell+1}^{+}(\alpha, 1)$	$I_{\ell+1}^{-}(\alpha, 1)$
$\beta_1 \leqslant \alpha \leqslant \beta_2$	$\alpha - \beta_1$	$\xi - \alpha$
$\beta_2 \leqslant \alpha \leqslant \xi$	$\alpha - \beta_1$	$\xi - \beta_2$

TABLE 8.2

	$I_{\ell+1}^+(\alpha, 2)$	$I_{\ell+1}^-(\alpha, 2)$
$\beta_2 \leqslant \alpha \leqslant \xi$	$\alpha - \beta_1$	$\xi - \beta_1$

It remains to determine α_ℓ^*, that is, to decide where to measure at the ℓth stage. This problem is attacked using dynamic programming. To this end, define the optimal value function

$$G_m(a_1, a_2) \underset{=}{\triangle} \begin{cases} \text{length of the final interval if the next } m \\ \text{observations are executed optimally, given} \\ \text{that the current relevant intervals are} \\ \text{of lengths } a_1 \text{ and } a_2, \text{ with } a_1 \geqslant a_2. \end{cases}$$

When $a_2 = 0$, the function φ_2 is deleted in later observations. Since for one function the optimal minimax procedure is *bisection*, $G_m(a_1, 0) = a_1/2^m$. It is clear that $G_m(\theta a_1, \theta a_2) = \theta G_m(a_1, a_2)$ for any $\theta > 0$. In particular $G_m(a_1, a_2) = a_1 G_m(1, a_2/a_1)$, so in discussing the general situation the right-hand side of the relevant intervals can be taken equal to one.

Let us assume that $m + 1$ observations are left and that the current relevant intervals are of lengths a_1 and a_2, as depicted in Figure 8.3. We already know, by (P2), that the next observation should be made on φ_1; let $\alpha = 1 - d$ be its position. Note that d must satisfy one of the two cases: case I ($0 \leqslant d \leqslant a_2$) and case II ($a_2 \leqslant d \leqslant a_1$). In each case there are two possible outcomes. Table 8.3 shows the resulting lengths of the relevant intervals (the longer written first).

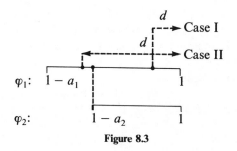

Figure 8.3

TABLE 8.3

	Case I	Case II
$\varphi_1(1 - d) > 0$	$a_1 - d; a_2 - d$	$a_1 - d; 0(\varphi_2 \text{ deleted})$
$\varphi_1(1 - d) < 0$	$a_2; d$	$d; a_2$

The optimality principle of dynamic programming, with the above observations, leads to the recursion formula

$$G_{m+1}(a_1, a_2) = \min \begin{cases} \min\limits_{0 \leqslant d \leqslant a_2} \max\left[G_m(a_1 - d, a_2 - d), G_m(a_2, d) \right] \\ \min\limits_{a_2 \leqslant d \leqslant a_1} \max\left[\dfrac{a_1 - d}{2^m}, G_m(d, a_2) \right]. \end{cases} \quad (8.4)$$

Formula (8.4) can be greatly simplified using the following lemma.

8.1 Lemma. Let h_1, h_2 be continuous decreasing functions and g_1, g_2 continuous increasing functions on the interval $[0, a_1]$ such that for $a_2 \in [0, a_1]$

$$h_1(a_2) = h_2(a_2); \qquad g_1(a_2) = g_2(a_2). \quad (8.5)$$

Suppose further that

$$g_1(0) \leqslant h_1(0); \qquad g_2(a_1) \geqslant h_2(a_1). \quad (8.6)$$

Then

$$\min\left\{ \min\limits_{0 \leqslant d \leqslant a_2} \max\left[h_1(d), g_1(d) \right]; \qquad \min\limits_{a_2 \leqslant d \leqslant a_1} \max\left[h_2(d), g_2(d) \right] \right\}$$

$$= \begin{cases} h_1(d_1) & \text{if } g_1(a_2) \geqslant h_1(a_2) \\ h_2(d_2) & \text{if } g_1(a_2) < h_1(a_2) \end{cases}$$

where d_1 and d_2 are solutions of

$$h_1(d_1) = g_1(d_1), \qquad h_2(d_2) = g_2(d_2).$$

∎

The proof can be readily deduced by inspecting Figure 8.4. Note that in case I the minimized function corresponds to the dashed line, and in case II to the asterisked line.

To use the lemma, substitute

$$h_1(d) = G_m(a_1 - d, a_2 - d), \qquad h_2(d) = \frac{a_1 - d}{2^m}$$

$$g_1(d) = G_m(a_2, d), \qquad g_2(d) = G_m(d, a_2)$$

and note that assumptions (8.5) and (8.6) hold. Therefore

$$G_{m+1}(a_1, a_2) = \begin{cases} G_m(a_1 - d_1, a_2 - d_1), & \text{if } G_m(a_2, a_2) \geqslant \dfrac{a_1 - d_1}{2^m} \\ \dfrac{a_1 - d_2}{2^m}, & \text{if } G_m(a_2, a_2) < \dfrac{a_1 - d_1}{2^m}, \end{cases} \quad (8.7)$$

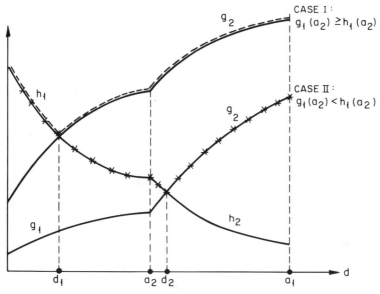

Figure 8.4

where d_1 and d_2 are the solutions of

$$G_m(a_1 - d_1, a_2 - d_1) = G_m(a_2, d_1)$$

$$\frac{a_1 - d_2}{2^m} = G_m(d_2, a_2).$$

A computer program based on formula (8.7) was used to graph $G_m(a_1, a_2)$ for different values of m. It revealed that for large m, G_m behaves linearly:

$$G_m \approx \gamma_m a_1 + \delta_m a_2.$$

This clue led to the following result.

8.2 Theorem.

$$G_m(a_1, a_2) \leqslant \frac{a_1 + u_m a_2}{2^m} \tag{8.8}$$

where u_m satisfies

$$u_0 = 0 \quad \text{and} \quad u_{m+1} = \frac{u_m^2 + u_m + 1}{u_m + \frac{1}{2}}. \tag{8.9}$$

Proof. The proof is by induction on m. For $m = 0$ both sides of the inequality (8.8) are equal to a_1. Suppose that Theorem 8.2 is valid for some (arbitrary) $m \geqslant 1$. It is shown below that Theorem 8.2 is also valid for $m + 1$. By (8.4), (8.8), and Lemma 8.1,

$$G_{m+1}(a_1, a_2) \leqslant \min \begin{cases} \min_{0 \leqslant d \leqslant a_2} \max\left[\dfrac{a_1 - d + u_m(a_2 - d)}{2^m}, \dfrac{a_2 + u_m d}{2^m} \right] \\[2em] \min_{a_2 \leqslant d \leqslant a_1} \max\left[\dfrac{a_1 - d}{2^m}, \dfrac{d + u_m a_2}{2^m} \right] \end{cases}$$

$$= \begin{cases} \dfrac{a_2 + u_m d_1}{2^m} & \text{if } a_1 < (2 + u_m)a_2 \\[1.5em] \dfrac{a_1 - d_2}{2^m} & \text{if } a_1 \geqslant (2 + u_m)a_2 \end{cases}$$

where

$$d_1 = \frac{a_1 + (u_m - 1)a_2}{2u_m + 1} \tag{8.10}$$

$$d_2 = \frac{a_1 - u_m a_2}{2}. \tag{8.11}$$

Substituting d_1 and d_2 into the above estimate, we have

$$G_{m+1}(a_1, a_2) \leqslant \begin{cases} \dfrac{1}{2^{m+1}} \left(\dfrac{u_m}{u_m + \frac{1}{2}} a_1 + \dfrac{u_m^2 + u_m + 1}{u_m + \frac{1}{2}} a_2 \right) & \text{if } a_1 < (2 + u_m)a_2 \\[1.5em] \dfrac{1}{2^{m+1}} (a_1 + u_m a_2) & \text{if } a_1 \geqslant (2 + u_m)a_2 \end{cases}$$

$$\leqslant \frac{a_1 + u_{m+1} a_2}{2^{m+1}}, \qquad \text{by (8.9).}$$

∎

In the proof of Theorem 8.2, a search method was constructed that produced the final interval of length at most

$$\frac{a_1 + u_m a_2}{2^m}.$$

The method is:

(MR2)

> Evaluate the sign of the function corresponding to the largest relevant interval at the point that is at distance d_1 from the right end point of this interval, if $a_1 \leqslant (2 + u_m)a_2$; otherwise at distance d_2 [where d_1 and d_2 are given by (8.10) and (8.11)].

We conclude that when solving Problem MINIROOT [for two functions by the method (MR2)], one obtains a final interval containing α^*, whose length $G_N(1, 1)$ does not exceed

$$E_N^2 \triangleq \frac{1 + u_N}{2^N}.$$

Here N is the number of observations and u_N is determined by (8.9). It can be shown that the error estimate E_N^2 asymptotically approaches $G_N(1, 1)$ (i.e., when $N \to \infty$). Thus (MR2) is "asymptotically optimal." In fact, it can be shown that for any N,

$$E_{N+2}^2 \leqslant G_N(1, 1) \leqslant E_N^2.$$

We now turn our attention to the general case of q functions. Following the discussion for the case $q = 2$, it can be shown that properties (P1) and (P2) remain valid. This leads to a natural extension of the recursion formula (8.4), which is described below.

At any stage of the search there are q relevant intervals with a common right end point with lengths

$$a_1 \geqslant a_2 \geqslant \cdots \geqslant a_q.$$

Let $G_m(a_1, a_2, \ldots, a_q)$ denote the length of the final interval if the next m observations are executed optimally. Then

$$G_{m+1}(a_1, a_2, \ldots, a_q) = \min_{j=1, 2, \ldots, q} \; \min_{a_{j+1} \leqslant d \leqslant a_j} R_j(d)$$

where

$$R_j(d) = \max\{ G_m(a_1 - d, a_2 - d, \ldots, a_j - d),$$

$$G_m(a_2, \ldots, a_j, d, a_{j+1}, \ldots, a_q)\}$$

and $a_{q+1} \triangleq 0$.

As is done earlier for Theorem 8.2, it can be shown that $G_m(a_1, a_2, \ldots, a_q)$ has an upper bound, which is a linear combination of the a_i's. This is proved by constructing a search method (MRq) generating a final interval (containing α^*) whose length does not exceed that upper bound.

8.3 Theorem

$$G_m(a_1, a_2, \ldots, a_q) \leqslant \frac{1}{2^m} \sum_{i=1}^{q} u_{m,i} a_i$$

where $u_{m,i}$ are defined recursively by

$$u_{m+1,i} = \max\left\{ \max_{i \leqslant j < i}\left(2 - \frac{u_{m,j}}{T_{m,j}}\right)u_{m,i},\right.$$

$$\left.\max_{i \leqslant j \leqslant q}\left[\left(2 - \frac{u_{m,j}}{T_{m,j}}\right)u_{m,i-1} + \frac{u_{m,j}}{T_{m,j}}u_{m,i}\right]\right\} \qquad (8.12)$$

where

$$T_{m,j} = u_{m,j} + \frac{1}{2}\sum_{l=1}^{j-1} u_{m,l}, \qquad m \geqslant 1, \qquad 2 \leqslant i \leqslant q \qquad (8.13)$$

with the initial conditions

$$u_{m,1} = 1 \qquad \forall \quad m \geqslant 0, u_{0,2} = u_{0,3} = \cdots = u_{0,q} = 0. \qquad (8.14)$$

The proof is patterned after the proof of Theorem 8.2 but is technically more involved, see GAL, BACHELIS, and BEN-TAL [78]; there it is also shown that for R_j^+ and R_j^- defined by

$$R_j^+ = \sum_{i=1}^{j} u_{m,i}(a_i - a_j)$$

$$R_j^- = u_{m,j}a_j + \sum_{i=2}^{j} u_{m,i-1}a_i + \sum_{i=j+1}^{q} u_{m,i}a_i, \qquad j = 1, 2, \ldots, q+1$$

there exists a unique index $j_0 \in \{1, 2, \ldots, q\}$ such that

$$R_{j_0}^+ < R_{j_0}^- \qquad \text{and} \qquad R_{j_0+1}^+ \geqslant R_{j_0+1}^- \qquad (8.15)$$

A description of the search method for solving the MINIROOT problem with q functions follows.

(MRq)

Order the relevant intervals by decreasing order $a_1 \geqslant a_2 \geqslant \cdots \geqslant a_q$. Evaluate the sign of the function corresponding to a_1 at the point that is at distance d from the right end point of the interval, where

$$d = \frac{1}{2T_{m,j_0}}\left[a_1 + \sum_{i=2}^{j_0}(u_{m,i} - u_{m,i-1})a_i - \sum_{i=j_0+1}^{q} u_{m,i}a_i\right].$$

Here j_0 is the index determined by (8.15), while T_{m,j_0} and the sequence $\{u_{m,i}\}$ are given in (8.12)–(8.14).

As is done earlier for (MR2), it can be shown that the method (MRq) is asymptotically optimal.

We emphasize that when solving Problem MINIROOT with q functions given N observations, the number

$$E_N^q = \frac{1}{2^N} \sum_{i=1}^{q} u_{N,i} a_i$$

is a *guaranteed* error estimate for the location of the minimal root α^*, *whatever the functions* $\varphi_1, \varphi_2, \ldots, \varphi_q$ *are.*

We conclude this section by illustrating the high efficiency of the (MRq) method in locating the minimal root of $q = 100$ functions (Table 8.4). Note, in particular, that when only 200 observations are allowed for locating the minimal root of 100 functions (i.e., "on the average" two observations per function) the (MRq) guarantees accuracy of at least $1.1 \cdot 10^{-7}$!

TABLE 8.4

Number of observations N	150	200	210	220	230	240	250
Accuracy E_N^{100}	$3.8 \cdot 10^{-2}$	$1.1 \cdot 10^{-7}$	$4.9 \cdot 10^{-9}$	$1.9 \cdot 10^{-10}$	$6.6 \cdot 10^{-12}$	$2.0 \cdot 10^{-13}$	$5.3 \cdot 10^{-15}$

SUGGESTED FURTHER READING

BAZARAA and SHETTY [79], BEN-TAL and ZLOBEC [75], GAL, BACHELIS, and BEN-TAL [78], HETTICH and BERGES [79], HIMMELBLAU [72], KARMANOV [75], LUENBERGER [73], MEYER [75], POLAK [71], SHEU [75], TOPKIS and VEINOTT [67], VANDERPLAATS and MOSES [76], ZANGWILL [67], ZOUTENDIJK [60 and 76].

3
SELECTED APPLICATIONS

In this chapter the theory developed so far is applied to three different problems: finding *Pareto optimal solutions*, solving a *lexicographic multicriteria problem*, and characterizing *Chebyshev solutions* of mathematical programs.

Each of these problems can be formulated as an ordinary mathematical program (P) having the special feature that Slater's condition cannot be satisfied. The optimality conditions thus involve the minimal index set of binding constraints $\mathcal{P}^=$, which in these applications has a specific meaning.

We do not elaborate on all forms of optimality conditions but instead restrict our discussion to dual characterizations for differentiable functions.

9 PARETO OPTIMIZATION

A fundamental concept in the theory of multicriteria decision making is that of the *Pareto optimum* (also called *efficient point, nondominated solution, admissible decision rule*).

9.1 Definition. Let $\{f^k : R^n \to R, k \in \mathcal{P}\}$ be a finite collection of functions ("criteria"). A point $x^* \in R^n$ is a *Pareto minimum* if there is no other point x such that

$$\begin{cases} f^k(x) \leqslant f^k(x^*), & k \in \mathcal{P} \\ \text{with at least one strict inequality.} \end{cases} \qquad (9.1)$$

Pareto optimality derives its name from the Italian economist PARETO, who in 1896 introduced the concept within the framework of welfare economics. Since then, Pareto optimality, in particular its relationships to competitive equilibrium, has been studied extensively by economists such as ARROW [51], KOOPMANS [51], and DEBREU [59]. The concept also plays an important role in statistical decision theory, especially with reference to optimal mixed strategies (see, for example, KARLIN [59] and FERGUSON [67]).

CHARNES and COOPER [61] made the observation that one can test the Pareto optimality of a point by solving a suitable mathematical program.

9.2 Proposition. A point x^* is a Pareto minimum for the criteria $\{f^k : k \in \mathcal{P}\}$ if and only if it is an optimal solution of the program

(MP)

$$\min \sum_{k \in \mathcal{P}} f^k(x)$$

s.t.

$$f^k(x) \leqslant f^k(x^*), \qquad k \in \mathcal{P}.$$

∎

Note that Slater's condition for (MP) is

"there exists $\bar{x}$ such that $f^k(\bar{x}) < f^k(x^*)$ for all $k \in \mathcal{P}$"

which obviously fails if x^* is indeed a Pareto optimum. Therefore characterizations, that assume Slater's condition, are not applicable, and one is compelled to use results such as those given in Chapter 1.

9.3 Theorem. Let $\{f^k : k \in \mathcal{P}\}$ be convex differentiable functions. A point x^* is a Pareto minimum if and only if for every proper subset Ω of $\mathcal{P}$ there exist non-negative numbers λ_k, $k \in \mathcal{P} \setminus \Omega$, not all zero, such that

$$\sum_{k \in \mathcal{P} \setminus \Omega} \lambda_k \nabla f^k(x^*) \in (D_\Omega^=(x))^*. \tag{9.2}$$

Proof. Applying to (MP) the dual form of the multi-Ω characterization (see Table 3.1), and noting that $\mathcal{P}(x^*) = \mathcal{P}$, we conclude that x^* is a Pareto minimum if and only if for every $\Omega \subset \mathcal{P}$ there exist $\bar{\lambda}_0 \geqslant 0$, $\bar{\lambda}_k \geqslant 0$, $k \in \mathcal{P} \setminus \Omega$ not all zero, such that

$$\bar{\lambda}_0 \sum_{k \in \mathcal{P}} \nabla f^k(x^*) + \sum_{k \in \mathcal{P} \setminus \Omega} \bar{\lambda}_k \nabla f^k(x^*) \in (D_\Omega^=(x^*))^*$$

or

$$\bar{\lambda}_0 \sum_{k \in \Omega} \nabla f^k(x^*) + \sum_{k \in \mathcal{P} \setminus \Omega} (\bar{\lambda}_0 + \bar{\lambda}_k) \nabla f^k(x^*) \in (D_\Omega^=(x^*))^*. \tag{9.3}$$

Now, since

$$D_{\Omega}^{=}(x^*) \subset \left\{ d : \nabla f^k(x^*)d = 0, k \in \Omega \right\}$$

it follows that

$$\left\{ d : \nabla f^k(x^*)d = 0, k \in \Omega \right\}^* \subset (D_{\Omega}^{=}(x^*))^*$$

by Lemma 2.2(f). In particular

$$\pm \bar{\lambda}_0 \sum_{k \in \Omega} \nabla f^k(x^*) \in (D_{\Omega}^{=}(x^*))^*.$$

This and (9.3) then imply that (9.2) holds with $\lambda_k = \bar{\lambda}_0 + \bar{\lambda}_k$. In order to prove sufficiency, let us note that the above inclusion, specified for $\Omega = \mathcal{P}$, implies

$$\bar{\lambda}_0 \sum_{k \in \mathcal{P}} \nabla f^k(x^*) \in (D_{\mathcal{P}}^{=}(x^*))^*,$$

which is exactly (9.3) for $\Omega = \mathcal{P}$. Hence only proper subsets $\Omega \subset \mathcal{P}$ need checking. The fact that (9.2) implies (9.3) is trivial: take $\bar{\lambda}_0 = 0$ and $\bar{\lambda}_k = \lambda_k$. ∎

An alternative characterization of Pareto optimality is based on the following simple observation.

9.4   Proposition.   A point x^* is a Pareto minimum if and only if

$$\left\{ x : f^k(x) \leqslant f^k(x^*), k \in \mathcal{P} \right\} = \left\{ x : f^k(x) = f^k(x^*), k \in \mathcal{P} \right\}$$

or, equivalently, $\mathcal{P} = \mathcal{P}^{=}$ [the minimal index set of binding constraints for the problem (MP)].

Algorithm 3.9 can now be used to characterize Pareto optimality in the following constructive manner.

9.5   Theorem.   Let $\{ f^k : k \in \mathcal{P} \}$ be differentiable convex functions. A point x^* is a Pareto minimum if and only if the algorithm below terminates at $\mathcal{P}^{=} = \mathcal{P}$.

Initialization.   Set $\Omega = \emptyset$.

Step 1   Solve the system

$$\sum_{k \in \mathcal{P} \setminus \Omega} \lambda_k \nabla f^k(x^*) \in (D_{\Omega}^{=}(x^*))^* \tag{9.4}$$

$$\lambda_k \geqslant 0, \qquad k \in \mathcal{P} \setminus \Omega, \qquad \sum_{k \in \mathcal{P} \setminus \Omega} \lambda_k = 1.$$

Step 2   If (9.4) has no solution, set $\mathcal{P}^= = \Omega$ and terminate (x^* is not Pareto optimal); otherwise continue.

Step 3   Set

$$\overline{\Omega} = \{ k \in \mathcal{P} \backslash \Omega : \lambda_k > 0 \}$$

$$\Omega_{\text{NEW}} = \Omega \cup \overline{\Omega}.$$

Step 4   If $\Omega_{\text{NEW}} = \mathcal{P}$, set $\mathcal{P}^= = \Omega_{\text{NEW}}$ and terminate (x^* is Pareto optimal); otherwise set $\Omega = \Omega_{\text{NEW}}$ and return to step 1.   ∎

Exercises and Examples

9.1   A sufficient condition for Pareto optimality (see also KARLIN [59]). A point x^* is a Pareto minimum for the convex differentiable criteria $\{ f^k : k \in \mathcal{P} \}$ if there exist positive "prices" $\lambda_k > 0$, $k \in \mathcal{P}$, such that

$$\sum_{k \in \mathcal{P}} \lambda_k \nabla f^k(x^*) = 0, \qquad (9.5)$$

that is, if x^* is an unconstrained minimum of $\sum_{k \in \mathcal{P}} \lambda_k f^k(x)$.

Proof.   The existence of positive λ_k's satisfying (9.5) guarantees that the algorithm in Theorem 9.5 will terminate in the first iteration at step 4, since in step 2, $\overline{\Omega} = \mathcal{P}$.   ∎

Note that the above is the Kuhn-Tucker sufficient condition applied to (MP).

Although a constraint qualification cannot hold for problem (MP), a reduction condition may still work. One such reduction condition is employed below.

9.2   If the criteria $\{ f^k : k \in \mathcal{P} \}$ are differentiable and strictly convex, then x^* is a Pareto minimum if and only if there exist non-negative "prices" $\lambda_k \geqslant 0$, $k \in \mathcal{P}$, not all zero, such that (9.5) holds.

Proof.   A straightforward application of Corollary 3.3 to (MP).   ∎

10   LEXICOGRAPHIC MULTICRITERIA PROGRAMS

In many practical problems, optimal policies are determined by making decisions successively. First the "most important" objective is met. From the solutions that meet this objective a smaller set is then chosen to satisfy an objective second in importance. If there is more than one solution, a third criterion can be applied, and so on. Such a problem is termed a *lexicographic multicriteria decision making problem* (see, for example, LEBEDEV, PODINOVSKI, and STYRIKOVIC [71] and PODINOVSKI [72]).

Mathematically, the lexicographic problem can be formulated as follows. Let $x = (x_1, \ldots, x_m)^T$ and $y = (y_1, \ldots, y_m)^T$ be two vectors in R^m. Then x is "lexicographically smaller than or equal" to y, in symbols $x \prec y$, if either x equals y or the first nonzero component of $x - y$ is negative. Let $f^1, \ldots, f^m$ be functions: $R^n \to R$ and let Y be a nonempty subset of R^n. Denote

$$f(x) = \left[f^1(x), \ldots, f^m(x) \right]^T.$$

The lexicographic problem is to determine $x^* \in Y$ satisfying

$$f(x^*) \prec f(x) \qquad \forall x \in Y.$$

Such a vector x^* is called a *lexicographic optimal solution*.

The set X^* of all optimal solutions of the lexicographic problem can be obtained by solving the following sequence of programs.

(P1) Determine

$$\alpha_1 \underline{\Delta} \min \{ f^1(x) : x \in Y \};$$

(P2) Determine

$$\alpha_2 \underline{\Delta} \min \{ f^2(x) : f^1(x) \leqslant \alpha_1, \qquad x \in Y \};$$

and so on for (P3), . . . , (P($m - 1$)). Finally, solve

(Pm)

$$\min \{ f^m(x) : f^1(x) \leqslant \alpha_1, f^2(x) \leqslant \alpha_2, \ldots, f^{m-1}(x) \leqslant \alpha_{m-1}, \qquad x \in Y \}.$$

The set X^* is equal to the set of all optimal solutions of (Pm).

Henceforth, for the sake of simplicity, we continue under the assumptions that the objective functions are faithfully convex and that the feasible set Y is polyhedral:

$$Y = \{ x : x^T a^i \leqslant \beta_i, \quad i \in \mathcal{L} \} \tag{10.1}$$

With the above Y, program (Pm) becomes

(Pm)

$$\min f^m(x)$$

s.t.

$$f^k(x) - \alpha_k \leqslant 0, \qquad k = 1, 2, \ldots, m - 1 \qquad (10.2)$$

$$x^T a^i - \beta^i \leqslant 0, \qquad i \in \mathcal{L}. \qquad (10.3)$$

Every feasible point of (Pm) must satisfy the constraints (10.2) with equality. On the other hand, since the remaining constraints are linear, it follows that the minimal set of (nonlinear) binding constraints is

$$\mathfrak{N}^= = \{1, 2, \ldots, m - 1\}.$$

Now the dual version of the "minimal subset" theorem from Table 3.1, applied to program (Pm), yields the following characterization of a lexicographic optimum.

10.1 Theorem. Consider the lexicographic problem with an ordered set of criteria

$$f^1(x), f^2(x), \ldots, f^m(x)$$

and the feasible set given by (10.1). Then a point $x^* \in R^n$ is a lexicographic minimum if and only if

(i) x^* is a feasible solution of (Pm);
(ii) there exist non-negative numbers μ_i, $i \in \mathcal{L}(x^*)$ such that

$$\nabla f^m(x^*) + \sum_{i \in \mathcal{L}(x^*)} \mu_i a^i \in \sum_{k=1}^{m-1} (D_k^=(x^*))^* \qquad (10.4)$$

where $\mathcal{L}(x^*)$ is the index set of those constraints in (10.3) that are binding at x^*. ∎

We used here the faithful convexity of the objective functions, in conjunction with Lemma 2.2(g), to justify the formula

$$(\cap D_k^=)^* = \sum (D_k^=)^*.$$

10.2 Remark. If for some ς, $1 \leqslant \varsigma < m$, program (P$\varsigma$) has a unique solution, say x^ς, then the feasible set of the next program (P, $\varsigma + 1$) is the singleton $\{x^\varsigma\}$. Thus there is no need to continue beyond (Pς). For a characterization of a unique optimal solution, see Ex. 3.21.

Exercises and Examples

10.1 Consider the lexicographic problem with two ordered objective functions

$$f^1 = -x_1 + x_3^2, \qquad f^2 = x_1 - x_2 + x_3$$

and linear constraints $Y = \{x \in R^3 : x_i \leqslant 0, \ i = 1, 2, 3\}$. To determine a lexicographic minimum, one first solves

(P1)

$$\min f^1 = -x_1 + x_3^2$$

s.t.

$$x_i \leqslant 0, \qquad i = 1, 2, 3$$

and finds an optimal solution $x_1 = 0$, $x_2 = -1$, $x_3 = 0$, and $\alpha_1 = 0$. Note that this is not a unique optimal solution. (In fact every point $(0, x_2, 0)^T$ with $x_2 \leqslant 0$ is a solution.) Thus we proceed to the second problem:

(P2)

$$\min f^2 = x_1 - x_2 + x_3$$

s.t.

$$f^1 - \alpha_1 = -x_1 + x_3^2 \leqslant 0$$
$$x_i \leqslant 0, \qquad i = 1, 2, 3.$$

To apply condition (10.4) we first note that $\mathfrak{N}^= = \{1\}$, $D_1^= = \{(0, d_2, 0)^T : d_2 \in R\}$ and $(D_1^=)^* = \{(d_1, 0, d_3)^T : d_1 \in R, \ d_3 \in R\}$. The general optimal solution of (P2) is of the form $x^* = (0, x_2^*, 0)^T$, $x_2^* \leqslant 0$. We claim that x_2^* cannot be negative. For if $x_2^* < 0$, then $\mathcal{L}(x^*) = \{1, 3\}$ and (10.4) becomes

$$\begin{bmatrix} 1 \\ -1 \\ 1 \end{bmatrix} + \mu_1 \begin{bmatrix} 1 \\ 0 \\ 0 \end{bmatrix} + \mu_3 \begin{bmatrix} 0 \\ 0 \\ 1 \end{bmatrix} = \begin{bmatrix} d_1 \\ 0 \\ d_3 \end{bmatrix}, \ \mu_1 \geqslant 0, \mu_3 \geqslant 0, d_1 \in R, d_3 \in R.$$

This system is obviously inconsistent.

11   CHEBYSHEV SOLUTIONS

For a system of linear equations $Ax = b$, a Chebyshev solution is one that minimizes the "l_∞-norm" of the residual $Ax - b$. For consistent systems of equations, the notion of a Chebyshev solution coincides with the ordinary solution.

A natural extension to mathematical programs (which may have no feasible solutions) is given in the following definition.

11.1   Definition.   Consider the program

(P)

$$\min f^0(x)$$

s.t.

$$f^k(x) \leqslant 0, \qquad k \in \mathcal{P}.$$

Associated with (P) is the program

(CP)

$$\min f^0(x)$$

s.t.

$$f^k(x) \leqslant a, \qquad k \in \mathcal{P}$$

where

$$a \underset{=}{\Delta} \max\left[0, \min_{x \in R^n} \max_{k \in \mathcal{P}} f^k(x)\right].$$

An optimal solution of (CP) is called a *Chebyshev solution* of (P). Note that if (P) has no feasible solution, then a is equal to $\hat{a}$, where

$$\hat{a} \underset{=}{\Delta} \min_{x \in R^n} \max_{k \in \mathcal{P}} f^k(x) > 0$$

[$\hat{a}$ is the minimal right-hand side for which (P) becomes feasible]. Moreover, if (P) *has* feasible solutions, then $a = 0$ and (CP) coincides with (P).

To compute $\hat{a}$, one solves the problem

$$\min \zeta$$

s.t.

$$f^k(x) - \zeta \leqslant 0, \qquad k \in \mathcal{P}. \tag{11.1}$$

Let $(\bar{x}, \bar{\zeta})$ be an optimal solution; then $\bar{x}$ is a Chebyshev solution and $\hat{a} = \bar{\zeta}$. Note that Slater's condition holds for (11.1). For later use we denote the binding constraints of (11.1) at $(\bar{x}, \bar{\zeta})$ by

$$B(\bar{x}) = \left\{ k \in \mathcal{P} : f^k(\bar{x}) = \bar{\zeta} \right\}.$$

Whenever the original program (P) is infeasible, the associated program (CP) never satisfies Slater's condition. Moreover the minimal set of binding constraints for (CP) is

$$\mathcal{P}^= = B(\bar{x}).$$

Therefore an immediate application of the dual optimality condition (from Table 3.1) yields the following theorem.

11.2   Theorem.   Let $\{ f^k : k \in \{0\} \cup \mathcal{P} \}$ be differentiable convex functions and x^* a feasible solution of (CP). Then x^* is a Chebyshev solution of (P) if and only if there exist non-negative numbers λ_k, $k \in \mathcal{P}(x^*) \backslash B(\bar{x})$

such that

$$\nabla f^0(x^*) + \sum_{k \in \mathscr{P}(x^*) \setminus B(\bar{x})} \lambda_k \nabla f^k(x^*) \in (D_{B(\bar{x})}^=(x^*))^*.$$

∎

Exercises and Examples

11.1 Let

$$f^1(x) = \begin{cases} x^2 + 2x + 2, & \text{if} & x \leqslant -1 \\ 1 & & -1 \leqslant x \leqslant 1 \\ x^2 - 2x + 2 & & x \geqslant 1 \end{cases}$$

and consider the program

$$\min f^0(x) = -x$$

s.t.

$$f^1(x) \leqslant 0$$

$$f^2(x) = x + 1 \leqslant 0.$$

Here an optimal solution of (11.1) is $\bar{x} = -\frac{1}{2}$, $\bar{\zeta} = 1$, and $B(\bar{x}) = \{1\}$. So $a = 1$ and (CP) is

$$\min f^0 = -x$$

s.t.

$$f^1 \leqslant 1$$

$$f^2 = x + 1 \leqslant 1.$$

We check the optimality of $x^* = 0$. Note that the constancy cone of f^1 at $x^* = 0$ is $D_1^=(x^*) = R$ (so its polar is zero) and $\mathscr{P}(x^*) = \{1, 2\}$. The optimality condition of Theorem 11.2 is

$$-1 + \lambda_2 \cdot 1 = 0, \qquad \lambda_2 \geqslant 0$$

which obviously holds with $\lambda_2 = 1$.

SUGGESTED FURTHER READING

BEN-ISRAEL, BEN-TAL, and CHARNES [77], BEN-TAL and ZLOBEC [77], CHARNES and COOPER [61], LEBEDEV, PODINOVSKI, and STYRIKOVIC [72], MAZUROV [72], PODINOVSKI and GAVRILOV [75], TIHONOV [66], ZUHOVITSKIY and AVDEYEVA [67].

PART 2

NONCONVEX PROGRAMMING

4

GENERAL NECESSARY
CONDITIONS

12 SECOND-ORDER DIRECTIONS

We start this chapter by introducing "primal elements," which are essential
in the formulation of the second-order necessary optimality conditions.

12.1 Definitions

(a) Let $f: R^n \to R$. A vector $d \in R^n$ is a *direction of decrease of f at*
x if there exist $\tau > 0$, a neighborhood N of d, and $\alpha > 0$, such
that for every $\bar{d} \in N$,

$$f(x + t\bar{d}) \leqslant f(x) - \alpha t \qquad \forall t \in [0, \tau].$$

The set of all such vectors (clearly an open cone) is denoted by
$K_f^<(x)$.

(b) A function $r: [0, \infty) \to R^n$ is called a *curve*. It is of *order* $o(t^k)$ if

$$\lim_{t \to 0} \frac{\|r(t)\|}{t^k} = 0 \qquad (\|\cdot\| \text{ is the Euclidean norm})$$

which is denoted by

$$r(t) \sim o(t^k).$$

(c) Let $H: R^n \to R^p$ be a mapping and $S \triangleq \{x : H(x) = 0\} \neq \emptyset$. A
vector d is a *tangent direction for S at* $x \in S$ if there exist $\tau > 0$
and a curve $r(t) \sim o(t)$ such that

$$x + td + r(t) \in S \qquad \forall t \in (0, T].$$

The set of such vectors (a cone) is denoted by $T_H(x)$. (The
above two types of cones have been extensively discussed in the
literature; see, for example, GIRSANOV [72].)

99

(d) For $f: R^n \rightarrow R$, a vector $d \in R^n$ is a *direction of quasidecrease of f at x*, if for every $\alpha > 0$ there exists $\tau > 0$ such that

$$f(x + td) \leqslant f(x) + \alpha t \qquad \forall t \in [0, \tau].$$

The set of such vectors (a cone) is denoted by $K_f(x)$. Clearly

$$K_f^<(x) \subset K_f(x).$$

(e) The set of *critical directions of f at x* is

$$K_f^=(x) \underset{=}{\Delta} K_f(x) \backslash K_f^<(x).$$

The following lemma lists several properties of directions of decrease. First we recall that a function $f: R^n \rightarrow R$ is *locally Lipschitzian at x* if there exist $\epsilon > 0$ and $\rho > 0$ such that

$$|f(x^1) - f(x^2)| \leqslant \rho \|x^1 - x^2\|$$

for all x^1 and x^2 satisfying $\|x^i - x\| \leqslant \epsilon$, $i = 1, 2$.

12.2 Lemma.

(a) Let $f: R^n \rightarrow R$ be locally Lipschitzian at x. If the directional derivative

$$f'(x, d) \underset{=}{\Delta} \lim_{t \rightarrow 0^+} \frac{f(x + td) - f(x)}{t} \qquad (12.1)$$

exists and is negative, then $d \in K_f^<(x)$. Conversely, if $d \in K_f^<(x)$ and the limit (12.1) exists, then $f'(x, d) < 0$.

(b) In particular, if $f'(x, \cdot)$ is convex, then $K_f^<(x)$ is a convex open cone.

(c) If f is convex, then

$$K_f^<(x) = \{d : f'(x, d) < 0\}.$$

Proof. See GIRSANOV [72, Theorems 7.1–7.4]. Part (a) is a special case of Theorem 12.5 below. ∎

A characterization of the cone of directions of quasidecrease is the content of the following lemma.

12.3 Lemma. If the directional derivative $f'(x, d)$ exists, then

$$K_f(x) = \{d : f'(x, d) \leqslant 0\}.$$

Proof. If $d \in K_f(x)$, then by definition

$$\lim_{t \to 0^+} \frac{f(x + td) - f(x)}{t} \leqslant \alpha \qquad \forall \alpha > 0.$$

Hence $f'(x, d) \leqslant 0$. Conversely, let $f'(x, d) \leqslant 0$. If this limit is equal to zero, then

$$f(x + td) - f(x) = o(t) < \alpha t \qquad \text{for any } \alpha > 0$$

$$\text{(and all } t > 0 \text{ sufficiently small).}$$

If the limit is negative, then

$$\frac{f(x + td) - f(x)}{t} \leqslant -\beta \qquad \text{for some } \beta > 0$$

and hence $f(x + td) - f(x) < \alpha t$ for any $\alpha > 0$. Therefore, in either case, $d \in K_f(x)$. ∎

12.4 Definition (Second-order elements). For a fixed vector x, a direction d in R^n, and a function $f: R^n \to R$, we call $z \in R^n$ a *second-order direction of decrease of f at* (x, d) if there exist $\tau > 0$, a neighborhood N of z, and $\beta > 0$ such that for every $\bar{z} \in N$

$$f\left(x + td + \tfrac{1}{2}t^2\bar{z}\right) \leqslant f(x) - \beta t^2 \qquad \forall t \in [0, \tau].$$

The (open) set of such vectors is denoted by $\mathcal{Q}_f(x, d)$. The function f is *$\mathcal{Q}$-regular at x* if

$$d \in K_f^=(x) \Rightarrow \mathcal{Q}_f(x, d) \quad \text{is a convex set.}$$

Clearly

$$\mathcal{Q}_f(x, 0) = K_f^<(x). \tag{12.2}$$

Properties of second-order directions of the decrease are studied in the next three results.

12.5 Theorem. Let $f: R^n \to R$ be locally Lipschitzian at x. If the limit

$$f''(x, d; z) \underset{\Delta}{\triangleq} \lim_{t \to 0^+} \frac{f\left(x + td + \tfrac{1}{2}t^2z\right) - f(x)}{t^2} \tag{12.3}$$

exists and is negative, then $z \in \mathcal{Q}_f(x, d)$. Conversely, if $z \in \mathcal{Q}_f(x, d)$ and the limit (12.3) exists, then $f''(x, d; z) < 0$.

Proof. Suppose that z satisfies $f''(x, d, z) < 0$. Then for some $\beta > 0$,

$$f''(x, d; z) < -4\beta. \tag{12.4}$$

Now, by (12.3), there exists $\tau > 0$ sufficiently small such that

$$f\left(x + td + \tfrac{1}{2}t^2 z\right) \leqslant f(x) - 2\beta t^2 \qquad \forall t \in (0, \hat{\tau}]. \qquad (12.5)$$

Let ϵ be the radius of the neighborhood in which f is Lipschitzian with constant ρ. Consider the neighborhood of z

$$N = \left\{ \bar{z} : \|\bar{z} - z\| \leqslant \frac{2\beta}{\rho} \right\} \qquad (12.6)$$

and define

$$\tau \triangleq \min\left\{ 1, \hat{\tau}, \frac{\epsilon}{\|d\| + \tfrac{1}{2}\|z\| + \beta/\rho} \right\}.$$

For fixed but arbitrary $\bar{z} \in N$, and any $t \in (0, \tau]$, take

$$x^1 \triangleq x + td + \tfrac{1}{2}t^2 \bar{z}.$$

Then

$$\|x^1 - x\| = \|td + \tfrac{1}{2}t^2 \bar{z}\| \leqslant t\|d\| + \tfrac{1}{2}t^2 \|\bar{z}\|$$

$$\leqslant t\left(\|d\| + \tfrac{1}{2}\|\bar{z} - z\| + \tfrac{1}{2}\|z\|\right), \qquad \text{since} \quad t \leqslant \tau \leqslant 1$$

$$\leqslant t\left(\|d\| + \frac{\beta}{\rho} + \tfrac{1}{2}\|z\|\right), \qquad \text{since} \quad \bar{z} \in N$$

$$\leqslant \epsilon, \qquad \text{by definition of } \tau.$$

Therefore, by the Lipschitz condition, for every $\bar{z} \in N$ and $t \in (0, \tau]$

$$f\left(x + td + \tfrac{1}{2}t^2 \bar{z}\right) \leqslant f\left(x + td + \tfrac{1}{2}t^2 z\right) + \rho\|\tfrac{1}{2}t^2 \bar{z} - \tfrac{1}{2}t^2 z\|$$

$$\leqslant f(x) - 2\beta t^2 + \tfrac{1}{2}\rho\left(\frac{2\beta}{\rho}\right)t^2, \qquad \text{by (12.5) and (12.6)}$$

$$= f(x) - \beta t^2,$$

which shows that $z \in \mathscr{Q}_f(x, d)$.

Conversely, if $z \in \mathscr{Q}_f(x, d)$, then for some $\tau > 0$, $\beta > 0$

$$f\left(x + td + \tfrac{1}{2}t^2 z\right) \leqslant f(x) - \beta t^2 \qquad \forall t \in (0, \tau].$$

Hence

$$\lim_{t \to 0^+} \frac{f\left(x + td + \tfrac{1}{2}t^2 z\right) - f(x)}{t^2} \leqslant -\beta < 0.$$

∎

12.6 Corollary. If f is twice continuously differentiable at x, and $d \in K_f^=(x)$, then

$$\mathcal{Q}_f(x,d) = \{z : \nabla f(x)^T z + d^T \nabla^2 f(x)d < 0\}, \tag{12.7}$$

showing, in particular, that f is $\mathcal{Q}$-regular at x.

Proof. Since $d \in K_f^=(x)$,

$$d \in K_f(x) \qquad \text{and} \qquad d \notin K_f^<(x).$$

The first relation implies $\nabla f(x)^T d \leqslant 0$ (by Lemma 12.3) and the second implies $\nabla f(x)^T d \not< 0$ [by Lemma 12.2(a)]. Hence

$$\nabla f(x)^T d = 0. \tag{12.8}$$

By Taylor's theorem, applied to $F(t) \underline{\Delta} f(x + td + \frac{1}{2}t^2 z)$ at $t = 0$,

$$f(x + td + \tfrac{1}{2}t^2 z) = f(x) + t\nabla f(x)^T d + \tfrac{1}{2}t^2\big[\nabla f(x)^T z + d^T \nabla^2 f(x)d\big] + o(t^2)$$

$$= f(x) + \tfrac{1}{2}t^2\big[\nabla f(x)^T z + d^T \nabla^2 f(x)d\big] + o(t^2)$$

by (12.8). Hence

$$\lim_{t \to 0^+} \frac{f(x + td + \tfrac{1}{2}t^2 z) - f(x)}{t^2} = \tfrac{1}{2}\big[\nabla f(x)^T z + d^T \nabla^2 f(x)d\big]. \tag{12.9}$$

The function f, being continuously differentiable, is Lipschitzian. Thus formula (12.7) follows by Theorem 12.5 and (12.9).

The $\mathcal{Q}$-regularity follows, since the right-hand side of (12.7) is a (open) half-space. ∎

12.7 Lemma.

  (a) If $f: R^n \to R$ is continuous, then
$$d \in K_f^<(x) \Rightarrow \mathcal{Q}_f(x,d) = R^n.$$

  (b) If f is locally Lipschitzian at x, then
$$d \in \text{comp}\, K_f(x) \Rightarrow \mathcal{Q}_f(x,d) = \emptyset.$$

Proof.

  (a) When $d \in K_f^<(x)$, then for some $\alpha > 0$, $\tau > 0$
$$f(x + td) \leqslant f(x) - \alpha t \qquad \forall t \in (0, \tau]$$
$$< f(x) - \alpha t^2 \qquad \forall t \in (0, \min\{\tau, 1\}].$$

So, for *arbitrary* $z \in R^n$

$$f\left(x + td + \tfrac{1}{2}t^2 z\right) \leqslant f(x) - \alpha t^2$$

for all t sufficiently small, showing $z \in \mathcal{Q}_f(x, d)$.

(b)   When $d \in \text{comp } K_f(x)$, then for some $\alpha > 0$

$$f(x + td) > f(x) + \alpha t \qquad \text{for } t > 0 \text{ small enough.} \tag{12.10}$$

Suppose that $\mathcal{Q}_f(x, d)$ is not empty and contains, say, z. Then for some $\beta > 0, \tau > 0$

$$f\left(x + td + \tfrac{1}{2}t^2 z\right) \leqslant f(x) - \beta t^2 \qquad \forall t \in [0, \tau].$$

This and (12.10) imply

$$f\left(x + td + \tfrac{1}{2}t^2 z\right) + \beta t^2 \leqslant f(x) \leqslant f(x + td) - \alpha t,$$

from which

$$\alpha t \leqslant f(x + td) - f\left(x + td + \tfrac{1}{2}t^2 z\right) - \beta t^2$$

$$\leqslant \rho t^2 \|z\| - \beta t^2$$

for all $t > 0$ small enough, by the Lipschitz condition (with constant 2ρ). Hence

$$\alpha t \leqslant (\rho \|z\| - \beta) t^2$$

which cannot happen for all $t > 0$ sufficiently small.   ■

The second-order element that extends the notion of tangent directions is defined next.

12.8   Definition.   Let $H : R^n \to R^p$ be a mapping, $S \triangleq \{x : H(x) = 0\} \neq \varnothing$ and $d \in R^n$. A vector z is a *second-order tangent direction for S at* (x, d) if there exist $\tau > 0$ and a curve $e(t) \sim o(t^2)$ such that

$$x + td + \tfrac{1}{2}t^2 z + e(t) \in S \qquad \forall t \in [0, \tau].$$

The set of such vectors is denoted by $V_H(x, d)$. The mapping H is *V-regular at* x if

$$d \in T_H(x) \Rightarrow V_H(x, d) \qquad \text{is a convex set.}$$

Clearly

$$V_H(x, 0) = T_H(x). \tag{12.11}$$

The set of second-order tangent directions is now characterized. First we introduce some notation: $h^i : R^n \to R$ is the ith component of H, which is thus given as the column

$$H(\cdot) = \begin{bmatrix} h^1(\cdot) \\ \vdots \\ h^p(\cdot) \end{bmatrix}.$$

The Jacobian ($p \times n$ matrix) of H at x is denoted by $\nabla H(x)$. The Hessian matrix of h^i at x is $\nabla^2 h^i(x)$ and

$$H''(x)(d,d) = \begin{bmatrix} d^T \nabla^2 h^1(x)d \\ \vdots \\ d^T \nabla^2 h^p(x)d \end{bmatrix}.$$

12.9   Theorem.   Let the mapping $H : R^n \to R^p$ be twice continuously differentiable at a point x satisfying $H(x) = 0$. Assume that the Jacobian matrix $\nabla H(x)$ has a full row rank [i.e., the range space of $\nabla H(x)$ is R^p]. Then for every $d \in T_H(x)$

$$V_H(x,d) = \{ z : \nabla H(x)z + H''(x)(d,d) = 0 \}. \qquad (12.12)$$

In particular, H is V-regular at x.

Proof.   Let d be a vector satisfying

$$\nabla H(x)d = 0. \qquad (12.13)$$

Denote by A the Jacobian matrix $\nabla H(x)$ and consider the function $G : R^p \times R \to R$,

$$G(y,t) \underset{=}{\triangle} H\left(x + td + \tfrac{1}{2}t^2 z + A^T y\right). \qquad (12.14)$$

The derivative of G with respect to y at $(0,0)$ is

$$G_y(0,0) = AA^T$$

which is nonsingular by the assumption on the rank of A. Therefore, the implicit function theorem, applied to $G(y,t) = 0$, guarantees the existence of a twice continuously differentiable function $y(t)$ such that

$$y(0) = 0 \qquad (12.15)$$

and

$$H\left(x + td + \tfrac{1}{2}t^2 z + A^T y(t)\right) \equiv 0 \qquad (12.16)$$

in a neighborhood of $t = 0$. Implicit differentiation of (12.16) at $t = 0$ yields

$$\nabla H(x)d + AA^T \dot{y}(0) = 0 \qquad \left(\text{where } \dot{y} = \frac{dy}{dt}\right).$$

Now (12.13) and the nonsingularity of AA^T imply

$$\dot{y}(0) = 0. \tag{12.17}$$

Differentiating (12.16) twice at $t = 0$ we have

$$\nabla H(x)z + H''(x)(d,d) + AA^T\ddot{y}(0) = 0 \qquad \left(\ddot{y} = \frac{d^2y}{dt^2} \right).$$

If z belongs to the set on the right-hand side of (12.12), it follows that

$$\ddot{y}(0) = 0. \tag{12.18}$$

From (12.15), (12.17), (12.18), and Taylor's theorem, we conclude that $y(t) \sim o(t^2)$. Similarly, for $e(t) \underset{\Delta}{=} A^Ty(t)$, $e(t) \sim o(t^2)$, which, together with (12.16), shows that $z \in V_H(x,d)$.

We show above that for every d satisfying (12.13),

$$\{ z : \nabla H(x)z + H''(x)(d,d) = 0 \} \subset V_H(x,d).$$

The reverse inclusion is a simple exercise. Thus for such d's (12.12) holds. In particular, it holds for $d = 0$, in which case (12.12) reduces by (12.11) to

$$T_H(x) = \{ z : \nabla H(x)z = 0 \}.$$

This means that the set of d's satisfying (12.13) is nothing else but the tangent cone $T_H(x)$. ∎

The last result parallels that of Lemma 12.7(b).

12.10 Lemma.   For any mapping H,

$$d \in \text{comp } T_H(x) \Rightarrow V_H(x,d) = \emptyset.$$

Proof.   When $d \notin T_H(x)$, then for every $\tau > \emptyset$ and every curve $r(t) \sim o(t)$ it follows that

$$H(x + \hat{t}d + r(\hat{t})) \neq 0 \qquad \text{for some} \quad \hat{t} \in [0, \tau]. \tag{12.19}$$

If $V_H(x,d)$ is not empty and contains z, then for some $\hat{\tau} > 0$ and some curve $e(t) \sim o(t^2)$:

$$H(x + td + \tfrac{1}{2}t^2z + e(t)) = 0 \qquad \forall t \in [0, \hat{\tau}].$$

This shows that, with the choice $r(t) = \tfrac{1}{2}t^2z + e(t) \sim o(t)$, (12.19) is contradicted. ∎

13 DUAL ELEMENTS: SUPPORT FUNCTIONS

In the first-order theory the dual optimality conditions are formulated in terms of the *dual set*. For second-order conditions the relevant notion is that of the *support function*.

13.1 Definition. Let S be a subset of R^n. The *support function of S* is the function $\delta^*(\cdot \mid S): R^n \to R$ defined by

$$\delta^*(y \mid S) \triangleq \sup_{x \in S} y^T x.$$

The *effective domain* of $\delta^*(\cdot \mid S)$ is

$$\Delta(S) \triangleq \{ y \in R^n : \delta^*(y \mid S) < \infty \}.$$

We use the convention that the supremum over the empty set is $-\infty$.

Immediate properties of the support function are given in the following lemma.

13.2 Lemma. Let S be a nonempty subset of R^n. Then

(a) $\delta^*(\cdot \mid S)$ is a positively homogeneous closed convex function and $\Delta(S)$ is a closed convex cone (recall that a function is closed if its epigraph is a closed set in R^{n+1});

(b) If S is a cone, then

$$\Delta(S) = -S^* \qquad \text{(the negative dual cone)}$$

and $\delta^*(\cdot \mid S) = \delta(\cdot \mid -S^*)$, where $\delta(\cdot \mid T)$ denotes the *indicator function* of T, that is,

$$\delta(x \mid T) = \begin{cases} 0 & \text{if } x \in T \\ \infty & \text{if } x \notin T; \end{cases}$$

(c) If $0 \in \mathrm{cl}\, S$ then

$$\delta^*(y \mid S) \geqslant 0 \qquad \forall y \in \Delta(S).$$

Moreover, if $0 \in \mathrm{int}\, S$, then

$$\delta^*(y \mid S) > 0 \qquad \forall 0 \neq y \in \Delta(S).$$

■

We now build up formulas for computing the support function of sets having a special structure.

13.3 Lemma. Let S be a convex subset of R^n with a nonempty interior, and M a subspace such that

$$(\mathrm{int}\, S) \cap M \neq \emptyset.$$

If $y \in \Delta(S \cap M)$, then

$$\delta^*(y \mid S \cap M) = \inf_{u, v} \{ \delta^*(u \mid S) + \delta^*(v \mid M) : u + v = y \} \qquad (13.1)$$

and, in fact, the infimum is attained.

Proof. The inequality "$\leqslant$" in (13.1) is obvious. We proceed to prove the reverse inequality. Fix

$$y \in \Delta(S \cap M)$$

and denote

$$\beta \underline{\Delta} \delta^*(y \mid S \cap M) < \infty. \tag{13.2}$$

Then

$$y^T x \leqslant \beta \qquad \forall x \in S \cap M. \tag{13.3}$$

By the extension theorem in Ex. 13.1 and by (13.3) it follows that there exists a vector u such that

$$u^T x = y^T x \qquad \forall x \in M \tag{13.4}$$

$$u^T x \leqslant \beta \qquad \forall x \in S.$$

Therefore

$$\delta^*(u \mid S) \leqslant \beta. \tag{13.5}$$

Now for $v \underline{\Delta} y - u$ it follows, by (13.4), that

$$\delta^*(v \mid M) = \sup_{x \in M} (y - u)^T x = 0. \tag{13.6}$$

Finally (13.5) and (13.6) show that

$$\beta \geqslant \delta^*(u \mid S) + \delta^*(v \mid M) \qquad (u + v = y).$$

∎

Lemma 13.3 can be useful in many ways. In the following it helps to derive the support function of the intersection of sets.

13.4   Lemma.   Let $S_1, S_2, \ldots, S_p$ be convex subsets of R^n, each having a nonempty interior such that

$$\bigcap_{i=1}^{p} \text{int } S_i \neq \varnothing. \tag{13.7}$$

If $y \in \Delta(\cap_{i=1}^{p} S_i)$, then

$$\delta^*\left(y \mid \bigcap_{i=1}^{p} S_i\right) = \inf_{y^1, \ldots, y^p} \left\{ \sum_{i=1}^{p} \delta^*(y^i \mid S_i) : \sum_{i=1}^{p} y^i = y \right\}$$

and, in fact, the infimum is attained for some $y^1, y^2, \ldots, y^p$.

Proof.   Let $\tilde{X}$ be the p-fold Cartesian product of R^n, each $\tilde{x} \in \tilde{X}$ is thus

$$\tilde{x} = \begin{bmatrix} x^1 \\ \vdots \\ x^p \end{bmatrix}, \qquad x^i \in R^n,$$

and the induced inner product in $\tilde{X}$ is

$$\tilde{x}^T\tilde{y} = \sum_{i=1}^{P} (x^i)^T y^i. \tag{13.8}$$

Denote

$$\tilde{S} \triangleq \{\tilde{x} : x^i \in S_i, \quad i = 1, 2, \ldots, p\} \tag{13.9}$$

$$\tilde{M} \triangleq \{\tilde{x} : x^1 = x^2 = \cdots = x^p\}. \tag{13.10}$$

Note that $\tilde{S}$ is a convex set in $\tilde{X}$ with int $\tilde{S} \neq \emptyset$ (since int $S_i \neq \emptyset$) and $\tilde{M}$ is a subspace in $\tilde{X}$. Further, by (13.7),

$$(\text{int } \tilde{S}) \cap \tilde{M} \neq \emptyset.$$

Lemma 13.3 is thus applicable, and it is readily deduced that

$$\delta^*(\tilde{y} \mid \tilde{S} \cap \tilde{M}) = \inf\{\delta^*(\tilde{u} \mid \tilde{S}) + \delta^*(\tilde{v} \mid \tilde{M}) : \tilde{u} + \tilde{v} = \tilde{y}\}. \tag{13.11}$$

By (13.8)–(13.10), the left-hand side of (13.11) is

$$\delta^*(\tilde{y} \mid \tilde{S} \cap \tilde{M}) = \sup\left\{ \sum_{i=1}^{P} (x^i)^T y^i : x^i \in S_i, \quad i = 1, 2, \ldots, p, \right.$$

$$\left. x^1 = x^2 = \cdots = x^p \right\}$$

$$= \delta^*\left(\sum_{i=1}^{P} y^i \mid \bigcap_{i=1}^{P} S_i \right). \tag{13.12}$$

The right-hand side is similarly interpreted.

$$\delta^*(\tilde{u} \mid \tilde{S}) = \sup\left\{ \sum_{i=1}^{P} (u^i)^T x^i : x^i \in S_i, \quad i = 1, 2, \ldots, p \right\}$$

$$= \sum_{i=1}^{P} \delta^*(u^i \mid S_i). \tag{13.13}$$

$$\delta^*(\tilde{v} \mid \tilde{M}) = \sup\left\{ \sum_{i=1}^{P} (v^i)^T x^i : x^1 = x^2 = \cdots = x^p \right\}$$

$$= \delta^*\left(\sum_{i=1}^{P} v^i \mid R^n \right)$$

$$= \begin{cases} 0 & \text{if } \sum_{i=1}^{P} v^i = 0 \\ \infty & \text{otherwise.} \end{cases} \tag{13.14}$$

Combining (3.11)–(3.14) we get

$$\delta^*\left(\sum_{i=1}^{p} y^i \mid \bigcap_{i=1}^{p} S_i\right) = \inf_{u^1, \ldots, u^p} \sum_{i=1}^{p} \delta^*(u^i \mid S_i).$$

■

Finally we compute the support function of the inverse set $A^{-1}S \underline{\underline{\Delta}} \{x \in R^n : Ax \in S\}$.

13.5 Lemma. Let A be an $m \times n$ matrix and S a convex set in R^m with nonempty interior such that

$$(\text{int } S) \cap R(A) \neq \emptyset. \tag{13.15}$$

If $y \in \Delta(A^{-1}S)$, then

$$\delta^*(y \mid A^{-1}S) = \inf_{u \in R^m} \{\delta^*(u \mid S) : A^T u = y\} \tag{13.16}$$

and the infimum is attained.

Proof. Let $\tilde{X} = R^n \times R^m$ and

$$S \underline{\underline{\Delta}} \left\{ \begin{bmatrix} x^1 \\ x^2 \end{bmatrix} \in \tilde{X} : x^2 \in S \right\}, \qquad \tilde{M} \underline{\underline{\Delta}} \left\{ \begin{bmatrix} x^1 \\ x^2 \end{bmatrix} \in \tilde{X} : Ax^1 = x^2 \right\}.$$

Note that $\tilde{S}$ is a convex set and $\text{int } \tilde{S} = R^n \times (\text{int } S) \neq \emptyset$. Also, $\tilde{M}$ is a subspace of $\tilde{X}$ and $\text{int } \tilde{S} \cap \tilde{M} \neq \emptyset$, by (13.15).

Choosing $\tilde{y} \in \tilde{X}$ to be of the form

$$\tilde{y} = \begin{bmatrix} y \\ 0 \end{bmatrix}, \qquad y \in R^n, \qquad 0 \in R^m$$

the left-hand side of (13.16) transforms into

$$\delta^*(y \mid A^{-1}S) = \sup\{ y^T x^1 : Ax^1 = x^2, \quad x^2 \in S \}$$
$$= \delta^*(\tilde{y} \mid \tilde{S} \cap \tilde{M}). \tag{13.17}$$

On the other hand, by Lemma 13.3

$$\delta^*(\tilde{y} \mid \tilde{S} \cap \tilde{M}) = \inf_{\tilde{u}, \tilde{v}} \{ \delta^*(\tilde{u} \mid \tilde{S}) + \delta^*(\tilde{v} \mid \tilde{M}) : \tilde{u} + \tilde{v} = \tilde{y} \}. \tag{13.18}$$

By the special form of $\tilde{y}$,

$$u^1 + v^1 = y, \qquad u^2 + v^2 = 0. \tag{13.19}$$

Each of the terms on the right-hand side of (13.18) is further computed:

$$\delta^*(\tilde{u} \mid \tilde{S}) = \sup_{x^1 \in R^n} (u^1)^T x^1 + \sup_{x^2 \in S} (u^2)^T x^2$$
$$= \begin{cases} \delta^*(u^2 \mid S) & \text{if } u^1 = 0 \\ \infty & \text{otherwise.} \end{cases} \tag{13.20}$$

To compute the second term, note that $\tilde{M}$ is the null space of the (partitioned) matrix $[A, -I]$, and its orthogonal complement $\tilde{M}^\perp$ is the range space of $[A, -I]^T$. Combining this with (13.19) implies

$$\tilde{v} = \begin{bmatrix} v^1 \\ -u^2 \end{bmatrix} \in \tilde{M}^\perp \Leftrightarrow v^1 = A^T u^2.$$

Then

$$\delta^*(\tilde{v} \mid \tilde{M}) = \delta(\tilde{v} \mid \tilde{M}^\perp), \qquad \text{by Lemma 13.2(b)}$$

$$= \begin{cases} 0 & \text{if } v^1 = A^T u^2 \\ \infty & \text{otherwise.} \end{cases} \tag{13.21}$$

Substituting (13.20) and (13.21) in (13.18), we have

$$\delta^*(\tilde{y} \mid \tilde{S} \cap \tilde{M}) = \inf_{u, v} \left\{ \delta^*(u^2 \mid S) : u^1 = 0, \quad v^1 = A^T u^2, \quad u^1 + v^1 = y \right\}$$

$$= \inf_{u^2} \left\{ \delta^*(u^2 \mid S) : A^T u^2 = y \right\}.$$

Comparison of this result with (13.17) proves (13.16). ∎

Exercises and Examples

AN EXTENSION THEOREM

The following result generalizes a result of Krein (see GIRSANOV [72, Theorem 5.1]) from convex *cones* to convex *sets*.

13.1 Let S be a convex set in R^n with nonempty interior and M a subspace of R^n such that $(\text{int } S) \cap M \neq \emptyset$. If y is a vector satisfying, for some scalar μ,

$$y^T x \leqslant \mu \qquad \forall x \in S \cap M$$

then there exists a vector u such that

$$u^T x = y^T x \qquad \forall x \in M$$
$$u^T x \leqslant \mu \qquad \forall x \in S.$$

Proof. Define the subset in R^{n+1}:

$$C \triangleq \left\{ \begin{bmatrix} s - m \\ -y^T m + \alpha \end{bmatrix} : s \in S, \quad m \in M, \quad \alpha \geqslant \mu \right\}.$$

It is easily verified that C is convex, $\text{int } C \neq \emptyset$ and $(0, 0)^T \notin \text{int } C$. Thus the standard separation theorem implies the existence of $(u, \beta) \neq (0, 0)$ such that

$$-u^T(s - m) + \beta(-y^Tm + \alpha) \geqslant 0 \qquad \forall m \in M, \qquad \forall s \in S, \qquad \forall \alpha \geqslant \mu.$$

Clearly $\beta \geqslant 0$ and, in fact, $\beta > 0$. (Otherwise $-u^T(S - M) \geqslant 0 \Rightarrow u = 0$.) Without loss of generality one can specify $\beta = 1$. Then

$$(u - y)^Tm + \mu \geqslant u^Ts \qquad \forall s \in S, \qquad \forall m \in M$$

and hence $u^TM = y^TM$ and $u^TS \leqslant \mu$. ∎

13.2 Let A be an $m \times n$ matrix and b a vector in R^m. Denote

$$S \underline{\Delta} \{x : Ax = b\}.$$

Then $\Delta(S) = R(A^T)$ and for every $y \in \Delta(S)$

$$\delta^*(y \mid S) = \min_u \{u^Tb : A^Tu = y\}. \tag{13.22}$$

Proof. Follows from elementary linear algebra. ∎

14 SECOND-ORDER NECESSARY CONDITIONS

The main tool in deriving second-order necessary conditions is the following generalization of the Dubovitskii-Milyutin separation Theorem 2.3.

14.1 Theorem. Let $S_0, S_1, \ldots, S_l$ be convex sets such that S_k, $k = 1$, $2, \ldots, l$ are open. Then the intersection

$$\bigcap_{k=0}^{l} S_k \tag{14.1}$$

is empty if and only if there are vectors

$$y^k \in \Delta(S_k), \qquad k = 0, 1, \ldots, l \tag{14.2}$$

not all zero, such that

$$\sum_{k=0}^{l} y^k = 0 \tag{14.3}$$

$$\sum_{k=0}^{l} \delta^*(y^k \mid S_k) \leqslant 0. \tag{14.4}$$

Proof. *If.* Suppose that some y^k's satisfy (14.2)–(14.4) but that $\cap S_k \neq \varnothing$; in particular assume that

$$x_0 \in \bigcap_{k=0}^{l} S_k.$$

Consider the sets $\hat{S}_k = S_k - x_0$, $k = 1, \ldots, l$. Then

$$0 \in \bigcap_{k=0}^{l} \hat{S}_k,$$

implying (see Lemma 13.2(c))

$$\delta^*(y_k | \hat{S}_k) \geqslant 0, \qquad k = 0, 1, \ldots, l$$

and, since S_k, $k = 1, \ldots, l$, are open,

$$\delta^*(y_j | \hat{S}_j) > 0, \qquad j = 1, \ldots, l, \qquad y_j \neq 0.$$

Indeed, for at least one $j = 1, \ldots, l$, say $j = j_0$, we have $y_{j_0} \neq 0$ [otherwise, by (14.3), $y_0 = y_1 = \cdots = y_l = 0$] and so

$$\delta^*(y_{j_0} | \hat{S}_j) > 0.$$

Then

$$\sum_{k=0}^{l} \delta^*(y_k | S_k) = \sum_{k=0}^{l} \delta^*(y_k | \hat{S}_k) + \sum_{k=0}^{l} x_0^T y^k.$$

The right-hand side equals zero, by (14.3), and the left-hand side is positive, by the above arguments, so a contradiction to (14.4) is obtained.

Only if. Let $S \triangleq \cap_{k=1}^{l} S_k$. Then S is open and (14.1) is equivalent to

$$S \cap S_0 = \emptyset. \tag{14.5}$$

We continue with the assumption that $S \neq \emptyset$. (This is the only interesting case. If $S = \emptyset$, apply the arguments below to a maximal nonvoid intersection of $S_1, S_2, \ldots, S_l$ and set $y^k = 0$ for k's that do not belong to this maximal intersection.) By the fundamental separation theorem (see, for example, LUENBERGER [69, Theorem 3, p. 133], GIRSANOV [72, Theorem 3.3]) S and S_0 can be separated by a hyperplane; in fact, (14.5) is equivalent to the existence of a vector $y \neq 0$ such that

$$\sup\{y^T x : x \in S\} \leqslant \inf\{y^T x : x \in S_0\}.$$

This inequality can be rewritten as

$$\delta^*\left(y | \bigcap_{k=1}^{l} S_k\right) + \delta^*(-y | S_0) \leqslant 0. \tag{14.6}$$

By Lemma 13.4, there exist $y^1, y^2, \ldots, y^l$ such that

$$\sum_{k=1}^{l} y^k = y \tag{14.7}$$

$$\delta^*\left(y | \bigcap_{k=1}^{l} S_k\right) = \sum_{k=1}^{l} \delta^*(y^k | S_k). \tag{14.8}$$

Setting $y_0 = -y$ it is seen that (14.7) is the same as (14.3) and, after (14.8) has been substituted in (14.6), that the latter is (14.4).   ∎

A different formulation of Theorem 14.1 is given in IOFFE and TIK-HOMIROV [68].

Consider the general nonlinear programming problem

(NLP)

$$\min f^0(x)$$

s.t.

$$f^i(x) \leqslant 0, \qquad i \in I \underline{\underline{\Delta}} \{1, 2, \ldots, m\}$$
$$H(x) = 0$$

where $f^i : R^n \to R$, $i \in \{0\} \cup I$ and $H : R^n \to R^p$ are continuous. For a feasible point x^* let

$$I(x^*) \underline{\underline{\Delta}} \{i \in I : f^i(x^*) = 0\}$$
$$I_0(x^*) \underline{\underline{\Delta}} \{0\} \cup I(x^*).$$

The cone $K_{f^i}(x^*)$ is abbreviated below as $K_i(x^*)$. Similar abbreviations are used for $K_{f^i}^{\leq}(x^*)$, $K_{f^i}^{=}(x^*)$, and $\mathfrak{Q}_{f^i}(x^*, d)$. Each vector d in $K_i(x^*)$ belongs to either $K_i^{<}(x^*)$ or $K_i^{=}(x^*)$. If

$$d \in \bigcap_{i \in I_0(x^*)} K_i(x^*)$$

the indices for which d is in $K_i^{=}(x^*)$ are collected in

$$J(x^*, d) \underline{\underline{\Delta}} \{i \in I_0(x^*) : d \in K_i^{=}(x^*)\}.$$

We are now in a position to formulate the second-order optimality condition.

14.2 Theorem. Let x^* be a local minimum for (NLP). Assume that the functions f^i, $i \in I_0(x^*)$ are $\mathfrak{Q}$-regular and that the mapping H is V-regular at x^*. Then to every $d \in R^n$ satisfying

$$\begin{cases} d \in K_i(x^*), & i \in I_0(x^*) \\ d \in T_H(x^*) \end{cases} \tag{14.9}$$

there correspond vectors

$$\begin{cases} y^i \in \Delta(\mathfrak{Q}_i(x^*, d)), & i \in J(x^*, d) \\ y \in \Delta(V_H(x^*, d)) \end{cases} \tag{14.10}$$

not all zero, which satisfy the *Euler-Lagrange equation*

$$y + \sum_{i \in J(x^*, d)} y^i = 0 \qquad (14.11)$$

and the *Legendre inequality*

$$\delta^*(y \mid V_H(x^*, d)) + \sum_{i \in J(x^*, d)} \delta^*(y^i \mid \mathcal{2}_i(x^*, d)) \leq 0. \qquad (14.12)$$

Proof. Let d be a fixed but arbitrary vector satisfying (14.9). First we show that there is no vector z such that

$$z \in \mathcal{2}_i(x^*, d) \qquad \forall i \in J(x^*, d) \qquad (14.13)$$

$$z \in V_H(x^*, d). \qquad (14.14)$$

Assume the contrary. The relations (14.13) imply the existence of neighborhoods N_i of z and scalars $\tau_i > 0$ such that for every $\bar{z} \in N_i$,

$$f^i(x^* + td + \tfrac{1}{2}t^2\bar{z}) < f^i(x^*) \qquad \forall t \in (0, \tau_i], \qquad i \in J(x^*, d). \qquad (14.15)$$

Similarly, (14.14) implies the existence of a curve $e(t) \sim o(t^2)$ and a scalar $\tau > 0$ such that

$$H(x^* + td + \tfrac{1}{2}t^2z + e(t)) = 0 \qquad \forall t \in [0, \tau]. \qquad (14.16)$$

Denote

$$\bar{z}(t) \underline{\Delta} z + \frac{e(t)}{t^2}.$$

Since $e(t) \sim o(t^2)$, it follows that for some sufficiently small $\hat{\tau} > 0$

$$\bar{z}(t) \in \bigcap_{i \in J(x^*, d)} N_i \qquad \forall t \in [0, \hat{\tau}]. \qquad (14.17)$$

For $0 \neq i \in J(x^*, d) \subset I_0(x^*)$, $f^i(x^*) = 0$ and it follows that for every t satisfying

$$0 < t < \bar{\tau} \underline{\Delta} \min\{\hat{\tau}, \tau, \min[\tau_i : i \in J(x^*, d)]\}:$$

$$f^i(x^* + td + \tfrac{1}{2}t^2\bar{z}(t)) < 0, \qquad 0 \neq i \in J(x^*, d) \qquad (14.18)$$

$$f^0(x^* + td + \tfrac{1}{2}t^2\bar{z}(t)) < f^0(x^*), \qquad \text{if } 0 \in J(x^*, d) \qquad (14.19)$$

by (14.15), (14.17), and also

$$H(x^* + td + \tfrac{1}{2}t^2\bar{z}(t)) = 0 \qquad (14.20)$$

by (14.16). Moreover if $0 \neq i \in I(x^*) \backslash J(x^*, d)$, then $d \in K_i^<(x^*)$, so that for some $\tau_i > 0$

$$f^i(x^* + td) < 0 \qquad \forall t \in [0, \tau_i]. \qquad (14.21)$$

Likewise, for some $\tau_0 > 0$

$$f^0(x^* + td) < f^0(x^*) \qquad \forall t \in (0, \tau_0], \qquad \text{if} \quad 0 \notin J(x^*, d). \quad (14.22)$$

Finally, if $i \notin I(x^*)$, then

$$f^i(x^*) < 0. \qquad (14.23)$$

Now (14.18)–(14.23) show that for all

$$0 < t \leqslant \tau^* \underline{\Delta} \min\left\{\bar{\tau}, \tau_0, \min\left[\tau_i : i \in I(x^*)\right]\right\}$$

the points on the curve

$$x^* + td + \tfrac{1}{2}t^2\bar{z}(t)$$

are feasible and give lower values for f^0 than x^*, thus contradicting the local optimality of x^*. This contradiction proves that a vector z satisfying (14.13)–(14.14) cannot exist, that is,

$$\bigcap_{i \in J(x^*, d)} \mathcal{Q}_i(x^*, d) \cap V_H(x^*, d) = \emptyset.$$

Note that (by $\mathcal{Q}$-regularity) each $\mathcal{Q}_i$ is an open convex set and that (by V-regularity) V_H is a convex set. So the separation theorem 14.1 applies and it establishes the existence of y^i, $i \in J(x^*, d)$ and y satisfying (14.10)–(14.12). ∎

14.3 Remark. In Theorem 14.2 the index set $J(x^*, d)$ can be replaced by $\{0\} \cup I$ with the *complementary slackness* type condition

$$y^i = 0 \qquad \text{if} \quad i \notin J(x^*, d)$$

added. Indeed, if $i \in I_0(x^*) \backslash J(x^*, d)$, then by Lemma 12.7(a) $\mathcal{Q}_i(x^*, d) = R^n$ and

$$\delta^*(u, \mathcal{Q}_i) = \begin{cases} 0 & \text{if} \quad u = 0 \\ \infty & \text{otherwise.} \end{cases}$$

This holds obviously also for $i \notin I_0(x^*)$.

14.4 Remark. The directions d *not* satisfying (14.9) do not add any new information. Indeed, if $d \notin K_i(x^*)$ for some $i \in I_0(x^*)$ [or $d \notin T_H(x^*)$], then by Lemma 12.7(b) [Lemma 12.10] $\mathcal{Q}_i(x^*, d) = \emptyset[V_H(x^*, d) = \emptyset]$. Hence for such d's the necessary condition holds trivially. Directions d that satisfy (14.9) are thus called *quasiusable directions*.

14.5 Remark. In the first-order conditions a constraint qualification assumption is needed to guarantee that the multiplier of the objective function is not zero. It is possible to state an assumption under which, for a given quasiusable direction d, the corresponding objective function multiplier $y^0 = y^0(d) \neq 0$.

Such an assumption is referred to as a *second-order constraint qualification at d*, denoted by CQ2(d).

An example of CQ2(d), for $d = \hat{d}$ satisfying $\hat{d} \in D_0^=(x^*)$, is

$$\bigcap_{0 \neq i \in J(x^*, \hat{d})} \mathcal{Q}_i(x^*, \hat{d}) \cap V_H(x^*, \hat{d}) \neq \emptyset. \tag{14.24}$$

Proof. Since $\hat{d} \in D_0^=(x^*)$, $0 \in J(x^*, \hat{d})$. Let $\hat{y}^i$, $\hat{y}$ be the vectors in Theorem 14.2 that correspond to the quasiusable direction $\hat{d}$. If these vectors satisfied (14.10)–(14.12) with $\hat{y}^0 = 0$, then (14.24) would be violated by Theorem 14.1. ∎

Note that for $\hat{d} = 0$, (14.24) reduces to

$$\bigcap_{i \in I(x^*)} K_i^<(x^*) \cap T_H(x^*) \neq \emptyset, \tag{14.25}$$

which is a (first-order) constraint qualification mentioned in GIRSANOV [72, Remark 3, p. 42]. In the differentiable case, (14.25) is the Mangasarian-Fromowitz constraint qualification.

Exercises and Examples

14.1 A first-order necessary condition: The Dubovitskii-Milyutin theorem. (See, for example, GIRSANOV [72, Theorem 6.1].) Let x^* be a local minimum for (NLP). Assume that the cones $K_i^<(x^*)$, $i \in I_0(x^*)$ and $T_H(x^*)$ are convex. Then

there exist $y^i \in \left[K_i^<(x^*) \right]^*$, $i \in I_0(x^*)$, and $y \in (T_H(x^*))^*$,

not all zero, such that

$$y + \sum_{i \in I_0(x^*)} y_i = 0.$$

Proof. This result follows from Theorem 14.2 by specializing to $d = 0$. Indeed, $d = 0$ satisfies (14.9) and $J(x^*, 0) = I_0(x^*)$. Also $\mathcal{Q}_i(x^*, 0) = K_i^<(x^*)$; $\mathcal{Q}$-regularity of $\mathcal{Q}_i$ is thus convexity of $K_i^<$ (see Definition 12.4). Similarly, $V_H(x^*, 0) = T_H(x^*)$ and V-regularity of V_H is convexity of T_H (see Definition 12.8). Moreover, by Lemma 13.2(b):

$$\Delta(\mathcal{Q}_i(x^*, 0)) = -(K_i^<(x^*))^*$$
$$\Delta(V_H(x^*, 0)) = -(T_H(x^*))^*$$
$$\delta^*(u \mid \mathcal{Q}_i(x^*, 0)) = \delta(u \mid -(K_i^<(x^*))^*)$$
$$\delta^*(u \mid V_H(x^*, 0)) = \delta(u \mid -(T_H(x^*))^*).$$

Substituting the above-calculated elements into Theorem 14.2 we obtain the Dubovitskii-Milyutin theorem.   ∎

14.2   Example for Theorem 14.2 (an adaptation of an example in FIACCO and McCORMICK [68, p. 24]).   Consider the nondifferentiable noncon-vex program

$$\min f^0 = (x_1 - 1)^2 + x_2^2$$

s.t.

$$f^1 = |x_1| - \frac{1}{k} x_2^2 \leqslant 0$$

where k is a positive parameter. For what values of k is $x^* = (0,0)^T$ a local minimum?

We calculate

$$\nabla f^0(x^*) = \begin{bmatrix} -2 \\ 0 \end{bmatrix}, \qquad \nabla^2 f^0(x^*) = \begin{bmatrix} 2 & 0 \\ 0 & 2 \end{bmatrix}$$

$$K_0^=(x^*) = \{ d : \nabla f^0(x^*)^T d = 0 \} = \left\{ \begin{bmatrix} 0 \\ d_2 \end{bmatrix} : d_2 \in R \right\}.$$

For $d \in K_0^=(x^*)$, $\mathcal{Q}_0(x^*,d) = \{ z : -2z_1 + 2d_2^2 < 0 \}$ (see Corollary 12.6); hence

$$\delta_0^*(y \mid \mathcal{Q}_0(x^*,d)) = \begin{cases} \lambda_0 d^T \nabla^2 f^0(x^*) d & \text{if } y = \lambda_0 \nabla f^0(x^*) \quad \text{for some } \lambda_0 \leqslant 0 \\ \infty & \text{otherwise} \end{cases}$$

$$= \begin{cases} 2\lambda_0 d_2^2 & \text{if } y = \lambda_0 \begin{bmatrix} -2 \\ 0 \end{bmatrix}, \qquad \lambda_0 \leqslant 0 \\ \infty & \text{otherwise.} \end{cases}$$

Further, by Lemma 12.3

$$K_1(x^*) = \{ d : (f^1)'(x^*,d) \leqslant 0 \}$$

$$= \left\{ \begin{bmatrix} d_1 \\ d_2 \end{bmatrix} : \lim_{t \to 0^+} \frac{t|d_1| - (1/k)t^2 d_2^2}{t} \leqslant 0 \right\}$$

$$= \left\{ \begin{bmatrix} 0 \\ d_2 \end{bmatrix} : d_2 \in R \right\}$$

and, by Lemma 12.2(a)

$$K_1^<(x^*) = \{ d : (f^1)'(x^*,d) < 0 \} = \emptyset;$$

so $K_1^=(x^*) = K_1(x^*)$.

Then for $d \in K_1^=(x^*)$, it follows by Theorem 12.5 that $\mathcal{Q}_1(x^*,d) =$

$\{z : (f^1)''(x^*, d; z) < 0\}$, which (after some manipulation) becomes

$$\mathcal{Q}_1(x^*, d) = \left\{ \begin{bmatrix} z_1 \\ z_2 \end{bmatrix} : \tfrac{1}{2}|z_1| < \tfrac{1}{k} d_2^2, \quad z_2 \in R \right\}$$

which is an open convex set. Finally, by direct computation,

$$\delta^*\left(y = \begin{bmatrix} y_1 \\ y_2 \end{bmatrix} \Big| \mathcal{Q}_1(x^*, d) \right) = \begin{cases} \tfrac{2}{k}|y_1|d_2^2 & \text{if } \quad y = \begin{bmatrix} y_1 \\ 0 \end{bmatrix}, \quad y_1 \in R \\ \infty & \text{otherwise.} \end{cases}$$

The Euler-Lagrange equation is then

$$\lambda_0 \begin{bmatrix} -2 \\ 0 \end{bmatrix} + \begin{bmatrix} y_1 \\ 0 \end{bmatrix} = \begin{bmatrix} 0 \\ 0 \end{bmatrix}, \quad \lambda_0 < 0, \quad y_1 \in R \qquad (14.26)$$

and the Legendre inequality is

$$2\lambda_0 d_2^2 + \tfrac{2}{k}|y_1|d_2^2 \leqslant 0, \quad d_2 \in R. \qquad (14.27)$$

From (14.26) it follows that $|y_1| = -2\lambda_0$, which when substituted in (14.27) yields

$$2\lambda_0 d_2^2\left(1 - \tfrac{2}{k}\right) \leqslant 0.$$

Hence $k \geqslant 2$, which answers the question.

In fact, it can be shown that $k \geqslant 2$ is also sufficient for optimality of x^*. Thus the second-order conditions of Theorem 14.2 here furnish complete information. Note that the first-order condition (see Ex. 14.1) does not provide any information, since $K_1^<(x^*) = \emptyset$ (for any $k > 0$) and therefore that condition holds trivially.

SUGGESTED FURTHER READING

BEN-TAL [77], BEN-TAL and ZOWE [79], GIRSANOV [72], HOLMES [75], IOFFE and TIKHOMIROV [68], LAURENT [72], PSHENICHNYI [71].

5

OPTIMALITY CONDITIONS FOR DIFFERENTIABLE PROGRAMS

15 NECESSARY CONDITIONS

Consider the problem

(DNLP)

$$\min f^0(x)$$

s.t.

$$f^i(x) \leqslant 0, \qquad i \in I \underline{\underline{\Delta}} \{1, 2, \ldots, m\}$$

$$h^j(x) = 0, \qquad j \in J \underline{\underline{\Delta}} \{1, 2, \ldots, p\}$$

where all the functions are from R^n to R and twice continuously differentiable. Let x^* be a feasible solution.

In the differentiable case the various sets and their support functions, appearing in the Euler-Lagrange equation and the Legendre inequality, can be given concrete formulas. These are now discussed.

First, for $d \in K_i^=(x^*)$:

$$\mathcal{Q}_i(x^*, d) = \left\{ z : \nabla f^i(x^*)^T z + d^T \nabla^2 f^i(x^*) d < 0 \right\} \qquad (15.1)$$

by Corollary 12.6. We apply Lemma 13.5, where we denote $A \underline{\underline{\Delta}} \nabla f^i(x^*)^T$ (a single-row matrix) and by S the open interval $(-\infty, -d^T \nabla^2 f^i(x^*) d)$. With this notation, (15.1) is rewritten as

$$\mathcal{Q}_i(x^*, d) = \{ z : Az \in S \}.$$

Therefore, whenever $\mathcal{Q}_i(x^*, d)$ is not empty, it follows by Lemma 13.5 that

$$\Delta[\mathcal{Q}_i(x^*, d)] = \{y^i : y^i = \lambda_i \nabla f^i(x^*) \qquad \text{for some} \quad \lambda_i \geqslant 0\} \quad (15.2)$$

and, for $y^i \in \Delta(\mathcal{Q}_i)$

$$\delta^*(y^i \mid \mathcal{Q}_i(x^*, d)) = -\lambda_i d^T \nabla^2 f^i(x^*) d. \qquad (15.3)$$

If the gradient vectors $\{\nabla h^j(x^*) : j \in J\}$ are linearly independent (that is, if the Jacobian

$$\nabla H(x^*) = \begin{bmatrix} \nabla h^1(x^*)^T \\ \vdots \\ \nabla h^p(x^*)^T \end{bmatrix}$$

has full row rank) it follows from Theorem 12.9 that

$$V_H(x^*, d) = \{z : \nabla h^j(x^*)^T z + d^T \nabla^2 h^j(x^*) d = 0 \qquad \forall j \in J\}.$$

Furthermore, by Ex. 13.2,

$$\Delta(V_H(x^*, d)) = \left\{y : y = \sum_{j \in J} \mu_j \nabla h^j(x^*) \qquad \text{for some} \quad \mu_j\text{'s}\right\} \quad (15.4)$$

and, for $y^i \in \Delta(V_H)$,

$$\delta^*(y \mid V_H(x^*, d)) = -\sum_{j \in J} \mu_j d^T \nabla^2 h^j(x^*) d. \qquad (15.5)$$

The set of quasiusable directions [i.e., directions d satisfying (14.9)] is here

$$D(x^*) \underline{\Delta} \left\{ d : \begin{array}{ll} \nabla f^i(x^*)^T d \leqslant 0, & i \in I_0(x^*) \\ \\ \nabla h^j(x^*)^T d = 0, & j \in J \end{array} \right\}.$$

Finally, the *Lagrangian function* $L : R^n \times R^{m+1} \times R^p \to R$ for (DNLP) is

$$L(x, \lambda, \mu) \underline{\Delta} \sum_{i=0}^{m} \lambda_i f^i(x) + \sum_{j=1}^{p} \mu_j h^j(x).$$

The above formulas are now readily used to derive a more concrete form of the general second-order conditions.

15.1   Theorem.   Let x^* be a local minimum for (DNLP). Then to every

$$d \in D(x^*)$$

there correspond vectors $\lambda = (\lambda_0, \lambda_1, \ldots, \lambda_m)^T$ and $\mu = (\mu_1, \mu_2, \ldots, \mu_p)^T$

$$\lambda_i \geqslant 0, \quad i = 0, 1, \ldots, m; \quad \mu_j \in R, \quad j = 1, 2, \ldots, p, \quad \text{not all zero} \quad (15.6)$$

such that

$$\nabla_x L(x^*, \lambda, \mu) = 0 \tag{15.7}$$

$$d^T \nabla_x^2 L(x^*, \lambda, \mu) d \geqslant 0 \tag{15.8}$$

$$\lambda_i f^i(x^*) = 0, \qquad i = 0, 1, \ldots, m \tag{15.9}$$

$$\lambda_i \nabla f^i(x^*)^T d = 0, \qquad i \in I_0(x^*). \tag{15.10}$$

Proof. We distinguish two cases.

Case I: $\{\nabla h^j(x^*) : j \in J\}$ are linearly dependent. Then there exist scalars $\bar{\mu}_j, j \in J$, not all zero, such that

$$\sum_{j \in J} \bar{\mu}_j \nabla h^j(x^*) = 0.$$

Hence (15.6)–(15.10) hold trivially with

$$\lambda = 0$$

$$\mu_j = \bar{\mu}_j \operatorname{sgn}\left[\sum_{l \in J} \bar{\mu}_l d^T \nabla^2 h^l(x^*) d\right].$$

Case II: Here $\{\nabla h^j(x^*) : j \in J\}$ are linearly independent. In this case formulas (15.4)–(15.5) are valid. Using these, together with formulas (15.2)–(15.3) in Theorem 14.2, and setting $\lambda_i = 0$ for $i \notin J(x^*, d)$, we prove the existence of λ and μ satisfying (15.6)–(15.10). ∎

15.2 Remarks

(a) The special case $d = 0$ in Theorem 15.1 yields the Mangasarian-Fromovitz condition (see MANGASARIAN and FROMO-VITZ [67]). It further reduces to the classical Fritz John condition for problems without equality constraints ($J = \emptyset$).

(b) For problems with equality constraints only ($I = \emptyset$), Theorem 15.1 coincides with the necessary condition in McSHANE [42].

Theorem 15.1 constitutes a departure from the classical second-order necessary conditions (see, for example, FIACCO and McCORMICK [68]), the main difference being the dependence of the multipliers (λ, μ) on the particular quasiusable direction d. Unless certain additional assumptions are imposed on the problem (and the solution x^*), the existence of a *fixed* multiplier vector [i.e., one that is the same for all $d \in D(x^*)$] cannot be guaranteed (see Ex. 15.1). Conditions that do guarantee a uniform behavior of the multipliers are thus termed *uniformity constraint qualifications* (UCQs).

A condition that CQ2(d) holds for every quasiusable direction d (see Remark 14.5) is called a *second-order constraint qualification* (CQ2).

Perhaps the simplest and most useful constraint qualification that is both UCQ and CQ2 is the following

"The vectors $\left\{ \nabla f^i(x^*), \nabla h^j(x^*) : i \in I(x^*), \quad j \in J \right\}$ are linearly independent." $\hspace{1cm}$ (15.11)

A feasible x^* satisfying (15.11) is called "regular" (see, for example, LUENBERGER [73]). In fact, any first-order constraint qualification (see PETERSON [73]) is a CQ2.

An example of a UCQ, which is not necessarily a CQ2, is the following condition:

(FMUCQ)
$$
\begin{aligned}
&\text{For every quasiusable direction } d \text{ there exists}\\
&z \in R^n \text{ such that}\\
&\nabla f^i(x^*)^T z + d^T \nabla^2 f^i(x^*) d = 0, \quad i \in I(x^*)\\
&\nabla h^j(x^*)^T z + d^T \nabla^2 h^j(x^*) d = 0, \quad j \in J.
\end{aligned}
$$

This condition is closely related to the "second-order Kuhn-Tucker constraint qualification" of FIACCO and McCORMICK [68].

In deriving the following result (known as the "second-order Kuhn-Tucker necessary condition" (see, for example, FIACCO and McCORMICK [68]) constraint qualifications are employed to achieve a considerable simplification of Theorem 15.1.

15.3 Corollary. Let x^* be a local minimum for (DNLP) and assume that either of the following holds:

(a) the linear independence condition (15.11),

(b) CQ2 together with FMUCQ.

Then there exist multipliers

$$\lambda_i \geqslant 0, \quad i = 1, 2, \ldots, m; \qquad \mu_j \in R, \quad j = 1, 2, \ldots, p \quad (15.12)$$

such that

$$\nabla f^0(x^*) + \sum_{i=1}^{m} \lambda_i \nabla f^i(x^*) + \sum_{j=1}^{p} \mu_j \nabla h^j(x^*) = 0, \quad (15.13)$$

$$\lambda_i f^i(x^*) = 0, \quad i = 1, 2, \ldots, m \quad (15.14)$$

and moreover, for every d satisfying

$$\begin{cases} \nabla f^i(x^*)^T d \leqslant 0, & i \in I(x^*) \\ \nabla h^j(x^*)^T d = 0, & j \in J \end{cases} \tag{15.15}$$

$$\lambda_i \nabla f^i(x^*)^T d = 0, \qquad i \in I(x^*) \tag{15.16}$$

it follows that

$$d^T \left\{ \nabla^2 f^0(x^*) + \sum_{i=1}^{m} \lambda_i \nabla^2 f^i(x^*) + \sum_{j=1}^{p} \mu_j \nabla^2 h^j(x^*) \right\} d \geqslant 0. \tag{15.17}$$

Proof. By Theorem 15.1, the optimality of x^* implies the existence of $\hat{\lambda}_i$'s and $\hat{\mu}_j$'s satisfying (15.6)–(15.10). Assume first that condition (a) holds. Then the multiplier $\hat{\lambda}_0$ must be positive, for otherwise (15.11) is violated. Thus

$$\lambda_0 = 1, \qquad \lambda_i = \frac{\hat{\lambda}_i}{\lambda_0}, \qquad i = 1, \ldots, m, \qquad \mu_j = \frac{\hat{\mu}_j}{\lambda_0}, \qquad j = 1, \ldots, p$$

$$\tag{15.18}$$

satisfy (15.12)–(15.14). Moreover, once the value one has been assigned to λ_0, the other multipliers are *uniquely* determined by (15.13). This, again, is a consequence of (15.11). Thus, obviously, the multipliers in (15.18) are the same for all $d \in D(x^*)$, and so they also satisfy (15.16) and (15.17) for every $d \in D(x^*)$. It remains to show that E, defined as the set of solutions of (15.15) and (15.16), here coincides with the set F of d's in $D(x^*)$ satisfying (15.10).

The inclusion

$$F \subset E$$

is obvious, since F contains two constraints more than E, namely,

$$\nabla f^0(x^*)^T d \leqslant 0 \qquad \text{and} \qquad \lambda_0 \nabla f^0(x^*)^T d = 0. \tag{15.19}$$

To prove that $E \subset F$, multiply (15.13) by d and use (15.15) and (15.16) to deduce that

$$\nabla f^0(x^*)^T d = 0.$$

It follows that (15.19) is also satisfied by every $d \in E$.

Under condition (b), for every quasiusable direction d the corresponding multiplier $\hat{\lambda}_0$ must be positive by the assumption CQ2. Hence the multipliers can be normalized as in (15.18) and these satisfy (15.12)–(15.14). Let d be some fixed quasiusable direction with the corresponding multipliers λ_i and μ_j (in Theorem 15.1). Let $\bar{d}$ be another quasiusable direction with corresponding multipliers $\bar{\lambda}_i$ and $\bar{\mu}_j$. Let $\bar{z}$ be the vector in (FMUCQ) associated with $\bar{d}$. All we have to show is that (15.16) and (15.17) also hold for $\bar{d}$ with λ_i, μ_j. Since $\lambda_0 = \bar{\lambda}_0 = 1$, it follows from (15.10) that

$$\nabla f^0(x^*)^T \bar{d} = 0$$

while from (15.17),

$$\nabla f^0(x^*) + \sum_{i=1}^{m} \lambda_i \nabla f^i(x^*) + \sum_{j=1}^{p} \mu_j \nabla h^j(x^*) = 0 \tag{15.20}$$

and

$$\nabla f^0(x^*) + \sum_{i=1}^{m} \bar{\lambda}_i \nabla f^i(x^*) + \sum_{j=1}^{p} \bar{\mu}_j \nabla h^j(x^*) = 0. \tag{15.21}$$

Multiplying (15.20) by $\bar{\alpha}$, and recalling that it is a quasiusable direction, yields

$$\lambda_i \nabla f^i(x^*)^T \bar{d} = 0, \qquad i \in I(x^*). \tag{15.22}$$

Moreover, by (15.18),

$$\bar{d}^T \left\{ \nabla^2 f^0(x^*) + \sum_{i=1}^{m} \lambda_i \nabla^2 f^i(x^*) + \sum_{j=1}^{p} \mu_j \nabla^2 h^j(x^*) \right\} \bar{d}$$

$$= \bar{d}^T \nabla^2 f^0(x^*) \bar{d} - \sum_{i=1}^{m} \lambda_i \nabla f^i(x^*)^T \bar{z} - \sum_{j=1}^{p} \mu_j \nabla h^j(x^*)^T \bar{z},$$

by (FMUCQ) and (15.9)

$$= \bar{d}^T \nabla^2 f^0(x^*) \bar{d} + \nabla f^0(x^*)^T \bar{z}, \qquad \text{by (15.20)}$$

$$= \bar{d}^T \nabla^2 f^0(x^*) \bar{d} - \sum_{i=1}^{m} \bar{\lambda}_i \nabla f^i(x^*)^T \bar{z} - \sum_{j=1}^{p} \bar{\mu}_j \nabla h^j(x^*)^T \bar{z},$$

by (15.21)

$$= \bar{d}^T \left\{ \nabla^2 f^0(x^*) + \sum_{i=1}^{m} \bar{\lambda}_i \nabla^2 f^i(x^*) + \sum_{j=1}^{p} \bar{\mu}_j \nabla^2 h^j(x^*) \right\} \bar{d},$$

again by (FMUCQ)

$$\geqslant 0, \qquad \text{since } \bar{d}, \bar{\lambda}_i, \bar{\mu}_j \text{ satisfy (15.8)}.$$

This inequality and (15.22) complete the proof. ∎

Theorem 15.1 can be given a primal version.

15.4 Corollary. Let x^* be a local minimum for (DNLP) and assume that the gradients $\{\nabla h^j(x^*): j \in J\}$ are linearly independent. Then for every $d \in D(x^*)$, the system

$$\begin{aligned} \nabla f^i(x^*)^T z + d^T \nabla^2 f^i(x^*)d < 0, \qquad & i \in J(x^*, d) \\ \nabla h^j(x^*)^T z + d^T \nabla^2 h^j(x^*)d = 0, \qquad & j \in J \end{aligned} \tag{15.23}$$

has no solution z.

Proof. Fix d in $D(x^*)$. Let $a = (a_i)$ and $b = (b_j)$ be the vectors with

$$\begin{aligned} a_i &= -d^T \nabla^2 f^i(x^*)d, \qquad i \in J(x^*, d) \\ b_j &= -d^T \nabla^2 h^j(x^*)d, \qquad j \in J; \end{aligned}$$

let A be a matrix whose ith column is $\nabla f^i(x^*)$, $i \in J(x^*, d)$ and B a matrix whose jth column is $\nabla h^j(x^*)$. Note that (15.9)–(15.10), with $\lambda_i = 0$, $i \notin J(x^*, d)$. Thus in the above notation Theorem 15.1 states that for every $d \in D(x^*)$ the system

$$A\lambda + B\mu = 0, \qquad a^T\lambda + b^T\mu \leqslant 0, \qquad 0 \neq \lambda \geqslant 0$$

has a solution λ, μ. (Note that the linear independence assumption guarantees $N(B) = \{0\}$, so $\lambda = 0$ would imply $\mu = 0$.) By the nonhomogenous alternative theorem (see Ex. 15.3), this is equivalent to the inconsistency of

$$A^T z < a, \qquad B^T z = b,$$

which is the system (15.23).   ∎

Exercises and Examples

15.1 Example for Theorem 15.1. Consider

$$\min f^0 = 2x_1 x_2 + \tfrac{1}{2}x_3^2$$

s.t.

$$f^1 = 2x_1 x_3 + \tfrac{1}{2}x_2^2 \leqslant 0$$
$$f^2 = 2x_2 x_3 + \tfrac{1}{2}x_1^2 \leqslant 0.$$

We show first that $x^* = (0,0,0)^T$ is a unique global minimum. If not, there is an $x \neq 0$ such that

$$f^i(x) \leqslant 0 \qquad i = 0, 1, 2. \tag{15.24}$$

The solution x cannot have a zero component, for otherwise, by (15.24), the two other components must be zero as well. Therefore, $x_i \neq 0$, $\forall i = 0, 1, 2,$

which again by (15.24) implies

$$x_1 x_2 < 0, \qquad x_1 x_3 < 0, \qquad x_2 x_3 < 0,$$

and hence $x_1^2 x_2^2 x_3^2 < 0$, a contradiction.

Next we demonstrate how the conditions of Theorem 15.1 are realized at $x^* = 0$. Note that $I(x^*) = \{1, 2\}$ and

$$\nabla f^i(x^*) = 0, \qquad i = 0, 1, 2.$$

Hence $D(x^*) = R^3$ and, moreover, the first-order conditions (15.7)–(15.10) are trivially satisfied. Further note that the second-order condition (15.8) is here

$$(d_1, d_2, d_3) \begin{bmatrix} \lambda_2 & 2\lambda_0 & 2\lambda_1 \\ 2\lambda_0 & \lambda_1 & 2\lambda_2 \\ 2\lambda_1 & 2\lambda_2 & \lambda_0 \end{bmatrix} \begin{bmatrix} d_1 \\ d_2 \\ d_3 \end{bmatrix} \geqslant 0. \tag{15.25}$$

Theorem 15.1 claims that for every $d \in R^3$ it is possible to find multipliers

$$\lambda_0 \geqslant 0, \qquad \lambda_1 \geqslant 0, \qquad \lambda_2 \geqslant 0, \qquad \text{not all zero} \tag{15.26}$$

such that (15.25) holds. Indeed, such multipliers are

$$(\lambda_0, \lambda_1, \lambda_2) = \begin{cases} (1,0,0) & \text{if} \quad d_1 d_2 \geqslant 0 \\ (0,1,0) & \text{if} \quad d_1 d_3 \geqslant 0 \\ (0,0,1) & \text{if} \quad d_2 d_3 \geqslant 0. \end{cases}$$

(The components of every $d \in R^3$ must satisfy at least one of the three inequalities: $d_i d_j \geqslant 0$, $\quad 3 \geqslant j > i \geqslant 1$.)

Let us note that it is impossible to find *fixed* multipliers satisfying (15.26) for which (15.25) holds for every $d \in R^3$. For then, after inserting in (15.25) the three vectors

$$\begin{bmatrix} 0 \\ 1 \\ -1 \end{bmatrix}, \begin{bmatrix} 1 \\ 0 \\ -1 \end{bmatrix}, \begin{bmatrix} -1 \\ 1 \\ 0 \end{bmatrix},$$

we obtain the inequalities

$$\lambda_0 + \lambda_1 - 4\lambda_2 \geqslant 0$$
$$\lambda_0 + \lambda_2 - 4\lambda_1 \geqslant 0$$
$$\lambda_1 + \lambda_2 - 4\lambda_0 \geqslant 0.$$

Adding them yields

$$-2(\lambda_0 + \lambda_1 + \lambda_2) \geqslant 0,$$

which contradicts (15.25).

15.2   A second-order constraint qualification.   An example of $CQ2(d)$ in the differentiable case is given below:

$$\begin{cases} \text{The gradients } \{ \nabla h^j(x^*) : j \in J \} \text{ are linearly independent and} \\ \text{there exists a vector } z \text{ satisfying} \\ \nabla f^i(x^*)^T z + d^T \nabla^2 f^i(x^*) d < 0, \quad i \in J(x^*, d) \\ \nabla h^j(x^*) z + d^T \nabla^2 h^j(x^*) d = 0, \quad j \in J. \end{cases} \tag{15.27}$$

By the discussion preceding Theorem 15.1 this condition is a realization of the general $CQ2(d)$ given in (14.24).

In example 15.1, (15.28) is satisfied, at the optimal solution $x^* = 0$, by $d = (-1, 1, 1)^T$ and arbitrary $z \in R^3$. It is indeed found there that $\lambda_0 \neq 0$.

For $d = 0$, (15.27) reduces to the so-called *modified Arrow-Hurwitz-Uzawa constraint qualification* (see MANGASARIAN and FROMOVITZ [67]).

15.3   A nonhomogeneous alternative theorem.   Let A and B be matrices, and a and b be vectors with appropriate dimensions. Assume that $b \in R(B^T)$. Then exactly one of the following two systems is consistent:

(i) $A\lambda + B\mu = 0, \quad a^T\lambda + b^T\mu \leqslant 0, \quad 0 \neq \lambda \geqslant 0,$
(ii) $A^T z < a, \quad B^T z = b.$

Proof.   System (ii) has no solution if and only if $\zeta^* \leqslant 0$ is the optimal value of the linear program

$$\max \zeta$$

s.t.

$$A^T z + \zeta e \leqslant a \tag{15.28}$$

$$B^T z = b, \quad \text{where } e \text{ is the vector of ones.}$$

The dual of (15.28) is

$$\min a^T\lambda + b^T\mu$$

s.t.

$$A\lambda + B\mu = 0$$

$$\lambda^T e = 1, \quad \lambda \geqslant 0.$$

Note that (15.28) is consistent since $b \in R(B^T)$. Thus the duality theorem applies; in particular it shows that $\zeta^* \leqslant 0$ if and only if system (i) is consistent.

16 SUFFICIENT CONDITIONS

We reconsider the problem (DNLP) and derive sufficient conditions for a feasible solution x^* to be an isolated local minimum. Recall that a point is an *isolated local minimum* if it is contained in a feasible neighborhood relative to which it is a unique minimum.

16.1 Theorem. Let x^* be a feasible point for (DNLP). Then x^* is an isolated local minimum if either $D(x^*) = \{0\}$, or for every nonzero $d \in D(x^*)$ there correspond vectors $\lambda = (\lambda_0, \lambda_1, \ldots, \lambda_m)^T$ and $\mu = (\mu_1, \mu_2, \ldots, \mu_p)^T$

$$\lambda_i \geqslant 0, \quad i = 0, 1, \ldots, m; \quad \mu_j \in R, \quad j = 1, 2, \ldots, p \quad \text{not all zero} \quad (16.1)$$

such that

$$\nabla_x L(x^*, \lambda, \mu) = 0 \tag{16.2}$$

$$d^T \nabla_x^2 L(x^*, \lambda, \mu) d > 0 \tag{16.3}$$

$$\lambda_i f^i(x^*) = 0, \quad i = 0, 1, \ldots, m \tag{16.4}$$

$$\lambda_i \nabla f^i(x^*) d = 0, \quad i \in I_0(x^*). \tag{16.5}$$

Proof. Suppose that x^* is not an isolated local minimum. Then there exists a sequence of feasible points $\{x^k\}, x^k \to x^*$ such that $f^0(x^k) \leqslant f^0(x^*)$ for all k sufficiently large. Specifically, $\{x^k\}$ can be chosen in the form $x^k = x^* + t_k d^k$, where $\{t_k\}$ is a sequence of positive scalars converging to zero, and $\{d^k\}$ is a sequence of normalized vectors ($\|d^k\| = 1$). For every $i \in I_0(x^*)$ we have $f^i(x^k) \leqslant f^i(x^*)$; hence by Taylor's expansion, for some $\xi_k^i \in [0, 1]$,

$$0 \geqslant f^i(x^* + t_k d^k) - f^i(x^*) = t_k \nabla f^i(x^*)^T d^k$$
$$+ \tfrac{1}{2} t_k^2 (d^k)^T \nabla^2 f^i(x^* + \xi_k^i t_k d^k) d^k.$$

After division by t_k:

$$0 \geqslant \nabla f^i(x^*)^T d^k + \tfrac{1}{2} t_k (d^k)^T \nabla^2 f^i(x^* + \xi_k^i t_k d^k) d^k, \quad i \in I_0(x^*). \tag{16.6}$$

Similarly, for the equality constraints:

$$0 = \nabla h^j(x^*)^T d^k + \tfrac{1}{2} t_k (d^k)^T \nabla^2 h^j(x^* + \eta_k^j t_k d^k) d^k, \quad j \in J \tag{16.7}$$

for some $\eta_k^j \in [0, 1], \quad j \in J$.

The limit of a convergent subsequence of $\{t_k, d^k\}$ (one that exists by compactness) is of the form $(0, \bar{d}), \|\bar{d}\| = 1$. Accordingly, (16.6) and (16.7) imply

$$\nabla f^i(x^*)^T \bar{d} \leqslant 0, \quad i \in I_0(x^*); \quad \nabla h^j(x^*)^T \bar{d} = 0, \quad j \in J$$

that is, $\bar{d} \in D(x^*)$, contradicting the first assumption $D(x^*) = \{0\}$. Let $\bar{\lambda}$, $\bar{\mu}$ be vectors that correspond to $\bar{d}$ and satisfy (16.1)–(16.5). Multiplying (16.6) by $\bar{\lambda}_i$ and (16.7) by $\bar{\mu}_j$, and summing up for all $i \in I_0(x^*)$ and $j \in J$, we have

$$0 \geqslant \tfrac{1}{2} t_k (d^k)^T \left\{ \sum_{i \in I_0(x^*)} \bar{\lambda}_i \nabla^2 f^i \left(x^* + \xi_k^i t_k d^k \right) + \sum_{j \in J} \bar{\mu}_j \nabla^2 h^j \left(x^* + \eta_k^j t_k d^k \right) \right\} d^k.$$

First dividing by $\tfrac{1}{2} t_k$ and then approaching the limit with the same convergent subsequence, we obtain

$$0 \geqslant \bar{d}^T \left\{ \sum_{i \in I_0(x^*)} \bar{\lambda}_i \nabla^2 f^i(x^*) + \sum_{j \in J} \bar{\mu}_j \nabla^2 h^j(x^*) \right\} \bar{d}.$$

This inequality contradicts (16.3)–(16.4). ∎

An equivalent formulation of Theorem 16.1 in a primal form is the content of the next result.

16.2 Corollary. Let x^* be a feasible point for (DNLP). Then x^* is an isolated local minimum if either $D(x^*)$ is empty or if for every nonzero $d \in D(x^*)$ the system

$$\begin{aligned} \nabla f^i(x^*)^T z + d^T \nabla^2 f^i(x^*) d \leqslant 0, \quad & i \in J(x^*, d) \\ \nabla h^j(x^*)^T z + d^T \nabla^2 h^j(x^*) d = 0, \quad & j \in J \end{aligned} \tag{16.8}$$

has no solution z.

Proof. The fact that $D(x^*) = \emptyset$ is a sufficient condition is shown earlier in the proof of Theorem 16.1. Assume now that $D(x^*) \neq \emptyset$. Let a, b, A, and B be as in the proof of Corollary 15.4. In terms of these, (16.8) is

$$A^T z \leqslant a, \quad B^T z = b. \tag{16.9}$$

We distinguish two cases.

Case I: Both a and b are equal to zero. Then the system

$$\begin{aligned} \nabla f^i(x^*)^T z \leqslant 0, \quad & i \in J(x^*, d) \\ \nabla h^j(x^*)^T z = 0, \quad & j \in J \end{aligned}$$

has no solution.

Case II: Here a and b are not both zero. The inconsistency of (16.9) is, by Ky Fan's theorem of the alternative (see Ex. 16.3), equivalent to the

TABLE 16.1. A Summary of Optimality Conditions in General Nonlinear Programming

Let x^* be a feasible solution of

$$(\text{DNLP}) \min \{f^0(x) : f^i(x) \le 0, \quad i \in I, h^j(x) = 0, \quad j \in J\}$$

where $\{f^i : i \in \{0\} \cup I\}$ and $\{h^j : j \in J\}$ are twice continuously differentiable functions: $R^n \to R$. Let $I(x^*) = \{i \in I : f^i(x^*) = 0\}$, $I_0(x^*) = \{0\} \cup I(x^*)$, $D(x^*) = \{d : \nabla f^i(x^*)^T d \le 0, i \in I_0(x^*), \nabla h^j(x^*)^T d = 0, j \in J\}$, and $J(x^*, d) = \{i \in I_0(x^*) : \nabla f^i(x^*)^T d = 0\}$.

Name and Type	Condition — Primal Statement	Condition — Dual Statement	Comments
Necessary condition for local minimum	For every $d \in D(x^*)$, there is no z such that $(z, d) \ne 0$ and $$\nabla f^i(x^*)^T z + d^T \nabla^2 f^i(x^*) d < 0, i \in J(x^*, d)$$ $$\nabla h^j(x^*)^T z + d^T \nabla^2 h^j(x^*) d = 0, j \in J.$$ The condition is valid if $\{\nabla h^j(x^*) : j \in J\}$ are linearly independent	For every $d \in D(x^*)$, there exist $\lambda_i > 0$, $i \in I$ and $\mu_j \in R, j \in J$, not all zero, such that $$\lambda_0 \nabla f^0(x^*) + \sum_{i \in I} \lambda_i \nabla f^i(x^*) + \sum_{j \in J} \mu_j \nabla h^j(x^*) = 0$$ $$d^T \left(\lambda_0 \nabla^2 f^0(x^*) + \sum_{i \in I} \lambda_i \nabla^2 f^i(x^*) + \sum_{j \in J} \mu_j \nabla^2 h^j(x^*) \right) d > 0$$ $$\lambda_i f^i(x^*) = 0, i \in I$$ $$\lambda_i \nabla f^i(x^*)^T d = 0, i \in I_0(x^*).$$	(i) For $d = 0$, the dual version reduces to the *Mangasarian-Fromovitz condition* (ii) If x^* is regular, or if the constraint qualifications CQ2 and FMUCQ both hold, then $\lambda_0 = 1$ and the multipliers λ_i, μ_j are fixed [the same for all $d \in D(x^*)$]. This is the so-called *second-order Kuhn-Tucker necessary condition*
Sufficient condition for isolated local minimum	Either $D(x^*) = \{0\}$ or, for every nonzero $d \in D(x^*)$, there is no z such that $(z, d) \ne 0$ and $$\nabla f^i(x^*)^T z + d^T \nabla^2 f^i(x^*) d < 0, j \in J(x^*, d)$$ $$\nabla h^j(x^*)^T z + d^T \nabla^2 h^j(x^*) d = 0, j \in J$$	Either $D(x^*) = \{0\}$, or for every nonzero $d \in D(x^*)$, there exist $\lambda_i > 0$, $i \in I$, and $\mu_j \in R, j \in J$, not all zero, such that (*) holds and $$\lambda_0 \nabla f^0(x^*) + \sum_{i \in I} \lambda_i \nabla f^i(x^*) + \sum_{j \in J} \mu_j \nabla h^j(x^*) = 0$$ $$d^T \left(\lambda_0 \nabla^2 f^0(x^*) + \sum_{i \in I} \lambda_i \nabla^2 f^i(x^*) + \sum_{j \in J} \mu_j \nabla^2 h^j(x^*) \right) d > 0.$$	The more restrictive case where $\lambda_0 = 1$ and the multipliers are fixed [the same for all $d \in D(x^*)$] is equivalent to the *Fiacco-McCormick second-order sufficient condition*

consistency of

$$A\lambda + B\mu = 0, \qquad a^T\lambda + b^T\mu < 0, \qquad \lambda \geqslant 0,$$

which is the system (16.1)–(16.5). ∎

The second-order necessary conditions and sufficient conditions, in primal and dual formulations for a differentiable nonlinear (nonconvex) program are summarized in Table 16.1.

Exercises and Examples

A SPECIAL CASE OF THEOREM 16.1

If the multipliers λ, μ are restricted to be fixed [i.e., the same for all $d \in D(x^*)$] with $\lambda_0 = 1$, Theorem 16.1 reduces to the following second-order sufficient condition of McCormick (see, for example, McCORMICK [67], FIACCO and McCORMICK [68]).

16.1 Let x^* be a feasible point for (DNLP). Then x^* is an isolated local minimum if there exist multipliers

$$\lambda_i \geqslant 0, \qquad i = 1, 2, \ldots, m; \qquad \mu_j \in R, \qquad j = 1, 2, \ldots, p$$

such that

$$\nabla f^0(x^*) + \sum_{i=1}^{m} \lambda_i \nabla f^i(x^*) + \sum_{j=1}^{p} \mu_j \nabla h^j(x^*) = 0$$

$$\lambda_i f^i(x^*) = 0, \qquad i = 1, 2, \ldots, m;$$

and moreover, for every d satisfying

$$\nabla f^i(x^*)^T d = 0 \qquad \text{if} \quad \lambda_i > 0$$

$$\nabla f^i(x^*)^T d \leqslant 0 \qquad \text{if} \quad \lambda_i = 0, \qquad i \in I(x^*)$$

$$\nabla h^j(x^*)^T d = 0, \qquad j = 1, 2, \ldots, p$$

it follows that

$$d^T \left\{ \nabla^2 f^0(x^*) + \sum_{i=1}^{m} \lambda_i \nabla^2 f^i(x^*) + \sum_{j=1}^{p} \mu_j \nabla^2 h^j(x^*) \right\} d \geqslant 0.$$

16.2 Example for Theorem 16.1. Consider again Ex. 15.1:

$$\min f^0 = 2x_1 x_2 + \tfrac{1}{2} x_3^2$$

s.t.

$$f^1 = 2x_1x_3 + \tfrac{1}{2}x_2^2 \leqslant 0$$

$$f^2 = 2x_2x_3 + \tfrac{1}{2}x_1^2 \leqslant 0.$$

There it was shown that $x^* = (0,0,0)^T$ is a unique (global) minimum. The sufficient condition of Theorem 16.1 is that to every $0 \neq d \in D(x^*) = R^3$, there correspond multipliers $\lambda_i \geqslant 0$, $i = 0, 1, 2$ not all zero, such that

$$(d_1\ d_2\ d_3)\begin{bmatrix} \lambda_2 & 2\lambda_0 & 2\lambda_1 \\ 2\lambda_0 & \lambda_1 & 2\lambda_2 \\ 2\lambda_1 & 2\lambda_2 & \lambda_0 \end{bmatrix}\begin{bmatrix} d_1 \\ d_2 \\ d_3 \end{bmatrix} > 0.$$

Indeed, the latter condition is satisfied by

$$(\lambda_0, \lambda_1, \lambda_2) = \begin{cases} (1,1,0) & \text{if} \quad d_1 = 0 \\ (1,0,1) & \text{if} \quad d_2 = 0 \\ (0,1,1) & \text{if} \quad d_3 = 0 \\ (1,0,0) & \text{if} \quad d_1 d_2 > 0 \\ (0,1,0) & \text{if} \quad d_1 d_3 > 0 \\ (0,0,1) & \text{if} \quad d_2 d_3 > 0. \end{cases}$$

Let us note that the sufficient condition (Ex. 16.1) with fixed multipliers requires that the matrix

$$\begin{bmatrix} \lambda_2 & 2\lambda_0 & 2\lambda_1 \\ 2\lambda_0 & \lambda_1 & 2\lambda_2 \\ 2\lambda_1 & 2\lambda_2 & \lambda_0 \end{bmatrix}$$

$(\lambda_0 = 1, \lambda_1 \geqslant 0, \lambda_2 \geqslant 0)$ be positive definite. But such a matrix is not even positive semidefinite, as is shown in Ex. 15.1.

16.3   Ky Fan's theorem of the alternative (FAN [56]). Let A and B be matrices and a and b vectors, not both zero, with appropriate dimensions. Then exactly one of the following two systems is consistent:

(i) $A\lambda + B\mu = 0, \qquad a^T\lambda + b^T\mu < 0, \qquad \lambda \geqslant 0$
(ii) $A^Tz \leqslant a, \qquad B^Tz = b.$

SUGGESTED FURTHER READING

BEN-TAL [77] and [80], BEN-TAL and ZOWE [79], BORWEIN [74], FIACCO and McCOR-MICK [68], GUIGNARD [69], HESTENES [75], McCORMICK [67], MAURER and ZOWE [79], ZLOBEC [71].

REFERENCES

ABRAMS, R. A.
[1975], "Projections of Convex Programs with Unattained Infima," *SIAM Journal on Control*, **13**, 706–718.

ABRAMS, R. A., and L. KERZNER
[1978], "A Simplified Test for Optimality," *Journal of Optimization Theory and Application*, **25**, 161–170.

ARROW, K. J.
[1951], "An Extension of the Basic Theorems of Classical Welfare Economics," *Proceedings of the Second Berkeley Symposium on Mathematical Statistics and Probability* (J. Neyman, Ed.), University of California Press, Berkeley, California.

AVRIEL, M.
[1976], *Nonlinear Programming: Analysis and Methods*, Prentice Hall, Englewoods Cliffs, N.J.

BAZARAA, M. S., and C. M. SHETTY
[1979], *Nonlinear Programming: Theory and Algorithms*, Wiley, New York.

BEN-ISRAEL, A.
[1969], "Linear Equations and Inequalities on Finite Dimensional, Real or Complex, Vector Spaces: A Unified Theory," *Journal of Mathematical Analysis and Applications*, **27**, 367–389.

BEN-ISRAEL, A., and A. BEN-TAL
[1976], "On a Characterization of Optimality in Convex Programming," *Mathematical Programming*, **11**, 81–88.

BEN-ISRAEL, A., A. BEN-TAL, and A. CHARNES
[1977], "Necessary and Sufficient Conditions for a Pareto Optimum in Convex Programming," *Econometrica*, **45**, 811–820.

BEN-ISRAEL, A., A. BEN-TAL, and S. ZLOBEC
[1979], "Optimality Conditions in Convex Programming," in *Survey of Mathematical Programming: Proceedings of the IX International Symposium on Mathematical Programming* (A. Prekopa, Ed.), Hungarian Academy of Sciences, Budapest, pp. 153–169.

BEN-TAL, A.
[1977], "Second Order Theory of Extremum Problems," Technical Report No. 107, Department of Computer Science, Technion-Israel Institute of Technology. Also in *Extremal Methods and Systems Analysis*, Lecture Notes in Economics and Mathematical Systems, No. 174, (A.V. Fiacco and K. O. Kortanek, Eds.), Springer-Verlag, New York, 1979.

BEN-TAL, A.
[1980], "Second Order and Related Extremality Conditions in Nonlinear Programming," *Journal of Optimization Theory and Application*, **31**, No. 2

135

BEN-TAL, A., and A. BEN-ISRAEL
[1976], "Primal Geometric Programs Treated by Linear Programming," *SIAM Journal of Applied Mathematics*, **30**, 538–556.

BEN-TAL, A., and A. BEN-ISRAEL
[1979], "Characterizations of Optimality in Convex Programming: The Nondifferentiable Case," *Applicable Analysis*, **9**, 137–156.

BEN-TAL, A., A. BEN-ISRAEL, and S. ZLOBEC
[1976], "Characterization of Optimality in Convex Programming Without a Constraint Qualification," *Journal of Optimization Theory and Applications*, **20**, 417–437.

BEN-TAL, A., and A. CHARNES
[1976], "Primal and Dual Optimality Criteria in Convex Programming," *Zeitschrift für Operations Research, Series Theory*, **21**, 197–209.

BEN-TAL, A., L. KERZNER, and S. ZLOBEC
[1980], "Optimality Conditions for Convex Semi-Infinite Programming Problems," *Naval Research Logistics Quarterly*, **27**, 413–435.

BEN-TAL, A., and S. ZLOBEC
[1975], "A New Class of Feasible Direction Methods," Research Report CCS216, Center for Cybernetic Studies, University of Texas, Austin. (Revised version: Technical Report TWISK61, National Research Institute for Mathematical Sciences, Pretoria, 1979.)

BEN-TAL, A., and S. ZLOBEC
[1977], "Convex Programming and the Lexicographic Multicriteria Problem," *Mathematische Operationsforchung und Statistik, Series Optimization*, **8**, 61–73.

BEN-TAL, A., and J. ZOWE
[1979], "A Unified Theory of First and Second Order Conditions for Extremum Problems in Topological Vector Spaces," Technical Report 156, Computer Science Department, Technion-Israel Institute of Technology, Haifa. *Mathematical Programming Studies* (forthcoming).

BORWEIN, J. M.
[1974], "Optimization with Respect to Partial Orderings," Ph.D. Thesis, Oxford University, Jesus College, Oxford, England.

BORWEIN, J. M., and H. WOLKOWICZ
[1979], "Characterization of Optimality without Constraint Qualification for the Abstract Convex Program," Department of Mathematics, Dalhousie University, Halifax, Nova Scotia, Canada.

CHARNES, A., and W. W. COOPER
[1961], *Management Models and Industrial Applications of Linear Programming*, Vol. 1, Wiley, New York.

CRAVEN, B., and S. ZLOBEC
[1978], "Complete Characterization of Optimality for Convex Programming Problems in Abstract Spaces," *Applicable Analysis* (forthcoming).

DEBREU, G.
[1959], *Theory of Value: An Axiomatic Analysis of Economic Equilibrium*, Wiley, New York.

DUBOVITSKII, A. Y., and A. A. MILYUTIN
[1963], "The Extremum Problem in the Presence of Constraints," *Doklady Akademiie Nauk SSSR*, **149**, 759–762 (in Russian).

DUFFIN, R. J., E. L. PETERSON, and C. ZENER
[1967], *Geometric Programming: Theory and Application*, Wiley, New York.

EREMIN, I. I., and N. N. ASTAFIEV
[1976], *Introduction to the Theory of Linear and Convex Programming*, Nauka, Moscow (in Russian).

FAN, K.
[1956], "On System of Linear Inequalities," in *Linear Inequalities and Related Systems* (H. W. Kuhn and A. W. Tucker, Eds.), Annals of Mathematics Studies, No. 38, Princeton University Press.

FERGUSON, T. S.
[1967], *Mathematical Statistics: A Decision Theoretic Approach*, Academic Press, New York.

FIACCO, A. V., and G. P. McCORMICK
[1968], *Nonlinear Programming: Sequential Unconstrained Minimization Techniques*, Wiley, New York.

GAL, S., B. BACHELIS, and A. BEN-TAL
[1978], "On Finding the Maximal Range of Validity of a Constrained System," *SIAM Journal of Control and Optimization*, **16**, 473–503.

GIRSANOV, I. V.
[1972], *Lectures on Mathematical Theory of Extremum Problems*, Lecture Notes in Economics and Mathematical Systems, No. 67, Springer-Verlag, New York.

GORDAN, P.
[1873], "Über die Auflosungen linearen Gleichungen mit reelen Coefficienten," *Mathematische Annalen*, **63**, 23–28.

GREENBERG, H. J., and W. P. PIERSKALLA
[1971], "A Review of Quasi-convex Functions," *Operations Research*, **19**, 1533–1570.

GUIGNARD, M.
[1969], "Generalized Kuhn-Tucker Conditions for Mathematical Programming Problems in a Banach Space," *SIAM Journal on Control*, **7**, 232–241.

HESTENES, M. R.
[1975], *Optimization Theory: The Finite Dimensional Case*, Wiley, New York.

HETTICH, R., and U. BERGES
[1979], "On Improving the Direction of Search of Steepest Descent Methods," presented at the IV Symposium on Operations Research, Saarbrücken, September 10–12, 1979.

HIMMELBLAU, D. M.
[1972], *Applied Nonlinear Programming*, McGraw-Hill, New York.

HOLMES, R. B.
[1975], *Geometric Functional Analysis and its Applications*, Springer-Verlag, New York.

IOFFE, A. D., and V. M. TIKHOMIROV
[1968], "Duality of Convex Functions and Extremum Problems," *Uspekhi Matematičeskii Nauk*, **23**, No. 6, 51–116. (English translation appeared in *Russian Mathematical Surveys*, **23**, No. 6, 53–123.)

JOHN, F.
[1948], "Extremum Problems with Inequalities as Subsidiary Conditions," in *Studies and Essays*, Courant anniversary volume (K. O. Friedrichs, O. E. Neugebauer, and J. J. Stoker, Eds.), Wiley-Interscience, New York.

KARAMARDIAN, A.
[1967], "Duality in Mathematical Programming," *Journal of Mathematical Analysis and Application*, **20**, 344–358.

KARLIN, S.
[1959], *Mathematical Methods and Theory in Games, Programming and Economics*, Vol. 1, Addison-Wesley, Reading, Massachusetts.

KARMANOV, V. G.
[1975], *Mathematical Programming*, Nauka, Moscow (in Russian).

KARUSH, W.
[1939], "Minima of Functions of Several Variables with Inequalities as Side Conditions," M.Sc. Thesis, Department of Mathematics, University of Chicago, Chicago, Illinois.

KOOPMANS, T. C.
[1951], *Activity Analysis of Production and Allocation* (T. C. Koopmans, Ed.), Wiley, New York, pp. 33–97.

KUHN, H. W.
[1976], "Nonlinear Programming: A Historial View," in *Nonlinear Programming* (R. W. Cottle and C. E. Lemke, Eds.), *SIAM-AMS Proceedings*, Vol. IX, Providence, Rhode Island.

KUHN, H. W., and A. W. TUCKER
[1951], "Nonlinear Programming," in *Proceedings of the Second Berkeley Symposium on Mathematical Statistics and Probability* (J. Neyman, Ed.), University of California Press, Berkeley, California.

LAURENT, J. P.
[1972], *Optimisation et Approximation*, Herman, Paris.

LEBEDEV, B. D., V. V. PODINOVSKI, and R. S. STYRIKOVIC
[1971], "An Optimization Problem with Respect to the Ordered Totality of Criteria," *Ekonomika i Matematičeskii Metody*, 7, 612–616.

LUENBERGER, D. G.
[1969], *Optimization by Vector Space Methods*, Wiley, New York.

LUENBERGER, D. G.
[1973], *Introduction to Linear and Nonlinear Programming*, Addison-Wesley, Reading, Massachusetts.

MANGASARIAN, O. L.
[1969], *Nonlinear Programming*, McGraw-Hill, New York.

MANGASARIAN, O. L., and S. FROMOWITZ
[1967], "The Fritz John Necessary Optimality Conditions in the Presence of Equality and Inequality Constraints," *Journal of Mathematical Analysis and its Applications*, 17, 37–47.

MARTOS, B.
[1975], *Nonlinear Programming, Theory and Methods*, North-Holland, Amsterdam.

MASSAM, H.
[1979], "Optimality Conditions for a Cone-Convex Programming Problem," *Journal of the Australian Mathematical Society (Series A)*, 27, 141–162.

MAURER, H., and J. ZOWE
[1979], "First and Second-Order Necessary and Sufficient Optimality Conditions for Infinite-Dimensional Programming Problems," *Mathematical Programming*, 16, 98–110.

MAZUROV, V. D.
[1972], "The Solution of an Ill-Posed Linear Optimization Problem under Contradictory Conditions," *Ekonomika i Matematičeskii Metody, Supplement*, Collection No. 3, 17–23 (in Russian).

McCORMICK, G. P.
[1967], "Second-Order Conditions for Constrained Minima," *SIAM Journal of Applied Mathematics*, 15, 641–652.

McSHANE, E. J.
[1942], "Sufficient Conditions for a Weak Relative Minimum in the Problem of Bolza," *Transactions of the American Mathematical Society*, 52, 344–379.

MEYER, R. R.
[1975], "Convergence Theory for a Class of Antijamming Strategies," MRC Technical Summary Report No. 1481, University of Wisconsin, Madison, Wisconsin.

MOND, B., and S. ZLOBEC
[1979], "Duality for Nondifferentiable Programming without a Constraint Qualification," *Utilitas Mathematica*, **15**, 291–302.

MOTZKIN, T. S.
[1936], "Beitrage zur Theorie der linearen Ungleichungen," Inaugural Dissertation, Basel, Jerusalem.

PARETO, V.
[1896], *Cours d'Economie Politique*, Rouge, Lausanne, Switzerland.

PETERSON, D. W.
[1973], "A Review of Constraint Qualifications in Finite-Dimension Spaces," *SIAM Review*, **15**, 639–654.

PODINOVSKI, V. V.
[1972], "Lexicographic Problems of Linear Programming," *Zurnal Vičislitel'noi Matematiki i Matematiceskoi Fiziki*, **12**, 1568–1571.

PODINOVSKI, V. V. and V. M. GAVRILOV
[1975], *Optimization with Respect to Successive Criteria*, Soviet Radio, Moscow (in Russian).

POLAK, E.
[1971], *Computational Methods in Optimization, a Unified Approach*, Academic Press, New York.

PONSTEIN, J.
[1967], "Seven Kinds of Convexity," *SIAM Review*, **9**, 115–119.

PSHENICHNYI, B. N.
[1971], *Necessary Conditions for an Extremum*, Dekker, New York.

ROBERTS, A. W. and D. E. VARBERG
[1973], *Convex Functions*, Academic Press, New York.

ROBINSON, S. M.
[1976], "First Order Conditions for General Nonlinear Optimization," *SIAM Journal of Applied Mathematics*, **30**, 597–607.

ROCKAFELLAR, R. T.
[1970a], *Convex Analysis*, Princeton University Press, Princeton, N.J.

ROCKAFELLAR, R. T.
[1970b], "Some Convex Programs Whose Duals are Linearly Constrained," *Nonlinear Programming* (J. B. Rosen, O. L. Mangasarian, and K. Ritter, Eds.), Academic Press, New York.

SANDBLOM, C.-L.
[1973], "A Computational Investigation into Nonlinear Decomposition," in *Decomposition of Large-Scale Problems* (D. H. Himmelblau, Ed.). North-Holland, Amsterdam, and American Elsevier, New York, pp. 415–426.

SHEU, C. Y.
[1975], "Optimal Elastic Design of Trusses by Feasible Direction Methods," *Journal of Optimization Theory and Applications*, **15**, 131–143.

STOER, J. and C. J. WITZGALL
[1970], *Convexity and Optimization in Finite Dimensions, I.*, Springer-Verlag, Berlin.

TIHONOV, P.
[1966], "On incorrectly posed problems of optimal planning," *Zurnal Vyčislitel'noi Matematiki i Matematičeskoi Fiziki*, **6**.

TOPKIS, D. M. and A. F. VEINOTT, Jr.
[1967], "On the Convergence of Feasible Direction Algorithms for Nonlinear Programming," *SIAM Journal of Control*, **5**, 268–279.

VANDERPLAATS, G. N. and F. MOSES
[1973], "Structural Optimization by Methods of Feasible Directions," *Computers and Structures*, **3**, 739–755.

WOLKOWICZ, H.
[1978], "Calculating the Cone of Directions of Constancy," *Journal of Optimization Theory and Applications*, **25**, 451–457.

WOLKOWICZ, H.
[1979], "Geometry of Optimality Conditions and Constraint Qualifications: The Convex Case," *Mathematical Programming*.

ZANGWILL, W. I.
[1967], *Nonlinear Programming: A Unified Approach*, Prentice-Hall, Englewood Cliffs, N.J.

ZLOBEC, S.
[1971], "Extensions of Asymptotic Kuhn-Tucker Conditions in Mathematical Programming," *SIAM Journal on Applied Mathematics*, **21**, 448–460.

ZLOBEC, S. and A. BEN-ISRAEL
[1979a], "Perturbed Convex Programs: Continuity of the Optimal Solutions and Optimal Values," in *Methods of Operations Research: Proceedings of the III Symposium on Operations Research* (W. Oettli and F. Steffens, Eds.), Verlag Athenäum/Hain/Scriptor/Hanstein, **31**, 737–749.

ZLOBEC, S., and A. BEN-ISRAEL
[1979b], "Duality in Convex Programming—a Linearization Approach," *Mathematische Operationsforschung und Statistik, Series Optimization*, **10**, 171–178.

ZLOBEC, S., R. GARDNER and A. BEN-ISRAEL
[1980], "Regions of Stability in Convex Programming," in *Proceedings of the I Symposium on Mathematical Programming with Data Perturbations* (A. V. Fiacco, Ed.), Dekker, New York.

ZLOBEC, S., and D. H. JACOBSON
[1979], "Minimizing an Arbitrary Function over a Convex Set," Technical Report TWISK93, National Research Institute for Mathematical Sciences, CSIR, Pretoria. Also in *Utilitas Mathematica* **16**, (1980). 291–302.

ZOUTENDIJK, G.
[1960], *Methods of Feasible Directions: A Study in Linear and Nonlinear Programming*, Elsevier, Amsterdam.

ZOUTENDIJK, G.
[1976], *Mathematical Programming Methods*, North-Holland, Amsterdam.

ZUKHOVITSKIY, S. I., and L. I. AVDEYEVA
[1967], *Linear and Convex Programming* (2nd ed.), Nauka, Moscow (in Russian).

GLOSSARY OF SYMBOLS

Symbol	description	introduced in page
R^n	n-dimensional Euclidean space	1
$\|x\|_1$	l_1-norm of x	54
$\|x\|_\infty$	l_∞-norm (Chebyshev norm) of x	94
dom f	domain of f	13
$f'(x, \cdot)$	the directional derivative of f at x	16
$\nabla f(x)$	the gradient (column vector) of f at x	17
$\nabla^2 f(x)$	the Hessian matrix of f at x	105
$D_f^<(x)$	cone of directions of descent	14
$D_f^{\leq}(x)$	cone of directions of nonascent	14
$D_f^=(x)$	cone of directions of constancy	14
$\delta^*(\cdot \mid S)$	the support function of S	107
$\Delta(S)$	the domain of $\delta^*(\cdot \mid S)$	107
cl(S)	the closure of S	26
$S \setminus T$	the set of elements belonging to S but not to T	19
ST	the Cartesian product of the sets S and T	55
conv(S)	the convex hull of S	14
conh	the conical hull of S	25
comp S	the complement of the set S	14
$\partial f(x)$	the subgradient set of f at x	27
$\varnothing$	the empty set	26
S^*	the dual set of S	25
$N(A)$	the null space of A	18
$R(A)$	the range space of A	24
$M^\perp$	the orthogonal complement of M	26
int S	the interior of S	35
card S	the cardinality of S	22
$\delta(\cdot \mid S)$	indicator function of S	107
$K_f^<(x)$	cone of directions of decrease	99
$T_H(x)$	tangent cone	99

$K_f(x)$	cone of directions of quasi-decrease	100
$K_f^=(x)$	set of critical directions	100
$r(t) \sim o(t^k)$	the curve $r(t)$ is of order $o(t^k)$	99
$\mathcal{D}_f(x,d)$	set of second-order directions of decrease of f at (x,d)	101
$V_H(x,d)$	set of second-order tangent directions of H at (x,d)	104
$\mathcal{P}(x^*)$	index set of binding constraints at x^*	19
$F(s,x^*)$	the set of feasible directions for s at x^*	19
$\mathcal{P}^=$	minimal index set of binding constraints	30
LP	linear programming	2
MP	mathematical programming	1
NLP	nonlinear programming	114
$DNLP$	differentiable NLP	120
CP	convex programming	1
NCP	nonconvex programming	2
$MELP$	method of elimination of linear programs	67
PFD	parametric feasible direction	70
MR_q	search method	86

INDEX